Matrix Transformations

$B = Q^{-1}AQ$ Similarity

$B = Q^{T}AQ$ Congruent

if $Q^{T} = Q^{-1}$ Orthogonal

$B = \bar{Q}^{T}AQ$

$= Q^{-1}AQ$ Unitary

$\bar{Q}^{T} = Q^{-1}$ "

LAMBDA-MATRICES AND VIBRATING SYSTEMS

LAMBDA-MATRICES AND VIBRATING SYSTEMS

Peter Lancaster
University of Calgary

Dover Publications, Inc.
Mineola, New York

Bibliographical Note

This Dover edition, first published in 2002, is an unabridged republication of the work first published as Volume 94 in the "International Series of Monographs in Pure and Applied Mathematics" by Pergamon Press, Oxford, in 1966. A new Preface to the Dover Edition and an errata list have been added to the present volume.

Library of Congress Cataloging-in-Publication Data

Lancaster, Peter, 1929–
Lambda-matrices and vibrating systems / Peter Lancaster. —Dover ed.
p. cm.
Originally published: Oxford : New York : Pergamon Press, [1966]
Includes bibliographical references and index.
ISBN 0-486-42546-0 (pbk.)
1. Vibration. 2. Matrices. I. Title.

QA871 .L27 2002
512.9'434—dc21

2002031424

Manufactured in the United States of America
Dover Publications, Inc., 31 East 2nd Street, Mineola, N.Y. 11501

TO MY PARENTS

with love and gratitude

CONTENTS

PREFACE TO THE DOVER EDITION

This edition of "Lambda-Matrices and Vibrating Systems" contains few changes from the original 1966 edition. Typographic slips and one or two mathematical inaccuracies have been corrected. Although the book has been out of print for many years, there has been a steady level of interest from the engineering community, as witnessed by literature references. More recently, there has been interest from the numerical analysis community. It has been recognized that quadratic eigenvalue problems (probably the book's major concern) are pervasive and have enough structure to justify the development of tailor-made algorithms taking advantage of modern software. Of course, this also requires some command of the spectral theory of lambda-matrices and hence the renewed interest in this book.

The theory of lambda-matrices (now more frequently known as matrix polynomials) has been extended a long way since 1966, and the place of this theory in a more general context of analytic and meromorphic matrix functions has been clarified. These developments have been stimulated and influenced by the demands of systems and control theory, which have evolved rapidly since 1966. A seminal work in this direction was the paper [GLR1] concerning self-adjoint problems (see the references at the end of this preface). This led to the volume [GLR2], which goes far beyond the theory of the first four chapters of the present volume. The reader interested in a deeper and more comprehensive theory is strongly recommended to consult that work. In particular, although the restriction of the present work to "simple" lambda-matrices gives a useful and convenient access to the theory it does, in fact, limit the usefulness of the theory.

The theory of [GLR2] evolved further with time and the reader will find other useful extensions of our topic in [GLR3] (especially for rational functions and self-adjoint problems) and in [GLR4] (for more general functions and for connections with systems theory).

Sections 3.6, 3.7 and 4.6 give the beginnings of factorization theory for lambda-matrices. This is an important topic which has also been extended considerably since 1966. Orthogonality properties of latent vectors (now more commonly eigen-vectors) discussed in Sections 4.4 and 4.7, when put into the context of self-adjoint problems, anticipates a later comprehensive theory depending on the structure of matrices acting on a linear space with an indefinite scalar product (see [GLR3], the reference to Krein and Langer [1] of this volume and subsequent work of H. Langer). The inverse problem discussed in Section 7.10 (expressing the coefficients in terms of spectral data) also has broad generalizations in [GLR2].

Some of the terminologies used in this volume now seem outdated, or even inappropriate; the definition of a "simple" λ-matrix, for example. Others include the definitions of *dimension* of a matrix, *order* of a square matrix, *regular* lambda-matrix, and *latent root*. No attempt has been made to change these usages. The reader may have to take care with the terminology, but the volume should remain internally consistent in this respect. It should be emphasized that "Lambda-Matrices and Vibrating Systems" still gives a viable introduction to the theory with a minimum of technical mathematical apparatus. This is the reason for its continuing popularity and for the publication of this new edition.

This revised edition has been prepared while the author visited the Institute of Mechanics of the Technical University of Darmstadt; a visit that was made possible by a Research Award of the Humboldt Foundation. It is fitting that this should be completed amongst a community of mechanical engineers (where the work of the first edition began), and the author is very grateful to the Humboldt Foundation and to Professor Peter Hagedorn for making this possible.

Darmstadt, January, 2002.

References

[GLR1] Gohberg, I., Lancaster, P. and Rodman, L., *Spectral analysis of self-adjoint matrix polynomials*, Annals of Math., **112** (1980), pp., 33–71.

[GLR2] Gohberg, I., Lancaster, P. and Rodman, L., *Matrix Polynomials*, Academic Press, 1982.

[GLR3] Gohberg, I., Lancaster, P. and Rodman, L., *Matrices and Indefinite Scalar Products*, Birkhäuser Verlag, 1983.

[GLR4] Gohberg, I., Lancaster, P. and Rodman, L., *Invariant Subspaces of Matrices with Applications*, John Wiley (Canadian Math. Soc. Monographs), 1986.

PAGE	LINE	CORRECTION
39	6	For c' read s'.
39	16	For "As we will in Chapter 7" read "It will be seen in Chapter 7 that"
40	3up	For "eqns." read "equations"
44	5	therefore
52	16	Insert comma after "hypothesis".
61	12	"...form, we obtain"
61	13	Fix braces.
64	12	The lowest "Q" should read "Q_β".
64	1up	An "l" is missing on the left. Thus: $R'_\alpha\{lA_0\lambda_1^{l-1} + \ldots$
65	17up	Space after the period. Space after "consider".
76	2	For "if he is" read "those who are".
76	3	For "then he might" read "should".
85	3	preceding
85	13-15	Italicize "the rates of...to compute."
87	8	seen
87	25	Insert after "...question" (they are generally fractals)
91	1up	Start new para with "The results..."
94	12	For i_s read μ_s.
100	10up	For "terms depending on t only" read "inhomogeneous terms"
105	12up	For "n-dimensional" read "ln-dimensional*". The star refers to a **footnote** to be added as follows: *This is a space of vector-valued functions of t.
106	1	Conversely,...
106	8-13	Replace two sentences "Observing...solutions of (6.2.3)." (approx. 5 lines) by:

Notice that each of these vectors has the structure of (6.2.5).

If $q_1^0(t), \ldots, q_{ln}^0(t)$ are linearly dependent, then there are numbers $\alpha_1, \ldots, \alpha_{ln}$, not all zero, such that

$$\sum_{j=0}^{ln} \alpha_j q_j^0(t) \equiv 0.$$

Differentiating successively $l-1$ times a contradiction is obtained with the linear independence of $q_1^+(t), \ldots, q_{ln}^+(t)$.

PAGE	LINE	CORRECTION
109	8up	Replace "allow of" by "admit".
109	4up	Replace "obvious" by "natural".
110	10	...general solution of (6.2.7) in the form...
112	8-9	...historically apt and, second, habitual...
123	1	...and the $2n$ solutions formed by the real and imaginary parts of $p_j = q_j e^{i\omega t}$ form...
123	3	...these $2n$...
123	4-5	...Notice also that these solutions represent simple harmonic motions in which...
125	8	For (7.4.3) read (7.4.8).
136	11	The statement of Theorem 7.9 should be changed as follows, and include reference to a footnote: "If all latent roots are nonzero* then $D(\lambda)$ is a simple λ-matrix." The footnote: "If C has rank $n-1$ then there is a defective zero latent root."
136	13	...a nonzero latent root...
137	12up	..is (complex) symmetric...
144	10	..is not always valid,...
144	1up	$e^{i(\omega t-\theta)}$
145	4	$e^{-i\theta}$
147	2	$e^{-i\theta_r}$
147	15	$e^{-i\theta_r}$
152	15,16up	Replace "which is much...at least." by "which may be more satisfactory for some purposes."
152	9up	...*if and only if* $\omega = \omega_s$ *for some* s.
153	12-13	Replace comma by period after equation (8.4.2) Replace line 12up by: This is the crux of the method: it is experimentally feasible to apply known in-phase forces, **k**, at a desired frequency ω and to measure directly the response $\mathbf{x}(\omega)$.
154	2	..understood that, as in the rest of this chapter, ω_s is...
169	14	For 9.4 read 9.5.
179	12-13	For "We fortunately...this problem in 2.9." read "Relevant information has been developed in 2.9."
179	16	For (9.1.6) read (9.1.1).
179	3up	Start new para with "For the perturbed..."

PAGE	LINE	CORRECTION
180	17	Start new para with "In order to..."
180	18	For "we now suppose that" read "suppose now that"
181	9up	It is found that...
184	6	...contains many results...
185	19	Space before "Stability..."
189	15	...iteration...

PREFACE

THE author's primary purpose in writing this work is to present under one cover several aspects and solutions of the problems of linear vibrating systems with a finite number of degrees of freedom, together with a careful account of that part of the theory of matrices required to deal with these problems efficiently. The treatment is intended to be mathematically sound and yet involve the reader in a minimum of mathematical abstraction. The results of the later chapters are then more readily available to those engaged in the practical analysis of vibrating systems. Indeed, the reader may prefer to dip straight into the later chapters on applications and refer back to the first four chapters for enlargement on theoretical issues as this becomes necessary. It should certainly be possible to start reading at Chapter 7 in this fashion.

While working on practical problems the author felt there was a considerable need for further investigations into the underlying problems concerning λ-matrices, or matrices whose elements are polynomials in the scalar λ. If complete mathematical solutions to the problems of λ-matrices were required, it soon became apparent that a great deal of sophisticated algebra would be involved, although treatment of the problems in full generality might, from the point of view of an applied mathematician, involve some wasted effort. Cases of only pathological interest were somehow to be avoided. Thus the interesting problem of where to make a division, if any, between a practically useful analysis and an all-embracing mathematically satisfying one then arose.

It has been claimed elsewhere that, in vibration theory, the cases in which repeated natural frequencies of vibration occur are of little physical significance, suggesting that one need consider only those cases in which all the natural frequencies are distinct. Unfortunately, experience shows that this is not enough. First, it commonly occurs that zero is a repeated natural frequency, and second, it occurs even more frequently that some natural frequencies of the system arise in one or more clusters. It has been found that understanding of the latter problems can only be achieved through study of problems in which natural frequencies are equal.

However, from an algebraic point of view the problems in which multiple proper values occur may be divided into two subclasses: those which do or do not have a full complement of proper vectors. Matrices for which each repeated proper value has a full complement of proper vectors are referred to in the literature as having linear elementary divisors, or as being non-derogatory, normalizable, or non-defective. In the author's experience, the physically significant problems are almost exclusively of this kind, and it is at this point that the line of practical utility has been drawn. It is claimed that, in the vast majority of practical vibration problems, any theory which assumes that a λ-matrix is simple may be applied without fear of its failing on this account.

In Chapter 1 we outline relevant portions of the theory of matrices. Theorems are proved where this seems appropriate, but we generally treat a fairly wide selection of topics briefly for future reference, and refer the reader to more thorough treatments in the vast literature on the subject. We include a statement of a theorem describing the Jordan canonical form which is, of course, essential for a complete understanding of the structure of square matrices. Though we do not hesitate to make use of the geometrical implications of this theorem in the theory of later chapters, a proof is avoided. We do so on the grounds that all the proofs are long and would distract us from our main objectives, or that they require more algebraic sophistication from the reader than we presuppose.

Chapters 2–4 then build up the theory of simple λ-matrices in a theorem-proof exposition. However, we continue to quote classical theorems without proof where convenient.

In Chapter 5, some new iterative numerical methods for λ-matrices are presented and rates of convergence are discussed. Although all the methods discussed are essentially for the improvement of known approximations, we nevertheless discuss the usefulness of the algorithms from the point of view of convergence in the large.

Chapter 6 presents some general solutions for simultaneous ordinary differential equations with constant coefficients. The analysis is facilitated by the use of results developed in Chapters 2–4. In Chapter 7 the results of the earlier chapters are applied to the vibration problems which stimulated the analysis. These vibration problems all involve simultaneous second order ordinary differential equations with constant coefficients. The results are presented in terms of the spectrum of the (in general) complex proper values and the modes of free vibration of the system.

An elegant theory of resonance testing is expounded in Chapter 8. The principle of the method appears to be due to Fraeijs de Veubeke and, in the author's view, deserves publication in a more permanent form. The method is presented here in the proper context of the theory of matrix pencils. It should be noted that all the results in Chapters 7 and 8 are exact, in the sense that no assumption is made about the magnitude of the damping forces in comparison with the inertial and elastic forces. Finally, Chapter 9 deals with some approximate methods for systems with viscous damping. Sections 9.2 and 9.3 are based on the theorems giving bounds for eigenvalues discussed in Chapter 1. Sections 9.4–9.7 contain an account of the application of perturbation theory to lightly damped systems.

It is a pleasure to acknowledge my debt to Prof. A. M. Ostrowski for guidance and stimulation provided on several occasions, Brian D. Warrack put a lot of time and effort into computer programming and operation to produce most of the results quoted in Chapter 5, and Lesia Hawrelak worked willingly and skilfully to produce the final typescript. I am grateful to them both and to several others for practical assistance given at various times. J. H. Wilkinson was good enough to provide data and his own solutions for the numerical example discussed in § 5. 10. Finally, I would like to record my gratitude to my wife, Edna, for encouragement and forbearance provided as the need arose.

P. LANCASTER
Pasadena
November, 1965

CHAPTER 1

A SKETCH OF SOME MATRIX THEORY

1.1 Definitions

In this chapter we do not aim to build up the theory of matrices as a logical structure. The intention is merely to pick out those ideas which are relevant to this work so that we may conveniently refer to them in later chapters, and at the same time to establish a system of notation and terminology. More detailed developments of most of the theory sketched in this chapter can be found in standard works by Aitken [1], Bellman [1], Ferrar [1], Frazer *et al.* [1], Gantmacher [1] and Mirsky [1], among many others. We will enter into detailed proofs or analysis where it is convenient to do so, and where the results are of particular importance in the development of succeeding chapters.

We define a *matrix* to be a rectangular array of numbers,

$$\begin{bmatrix} a_{11} & a_{12} \cdots & a_{1n} \\ a_{21} & a_{22} \cdots & a_{2n} \\ \vdots & \vdots & \vdots \\ a_{m1} & a_{m2} & a_{mn} \end{bmatrix}$$

where the *elements* of the matrix a_{ij} ($1 \leqq i \leqq m$, $1 \leqq j \leqq n$) are numbers from a field F. In this work the field F will either be that of the real numbers or that of the complex numbers. Matrices with more than one row and more than one column will generally be denoted by capital Roman letters. The *dimensions* of the matrix in terms of the number of rows, m, and the number of columns, n, are indicated by $m \times n$. When $m = n$ we say that the matrix is square of *order* n. In the general case the matrix is said to be *rectangular.*

From the $m \times n$ matrix A with elements a_{ij} we can define an $n \times m$ matrix with elements a_{ji}. This is known as the *transpose* of A and is denoted by A'.

We now define the following operations:

(i) The sum of $m \times n$ matrices A and B with elements a_{ij} and b_{ij} is the matrix C with elements $c_{ij} = a_{ij} + b_{ij}$, and we write $C = A + B$.

(ii) The product of a matrix A with a number α of the field F is the matrix C with elements given by $c_{ij} = \alpha a_{ij}$, and we write $C = \alpha A$.

(iii) The product of the $m \times l$ matrix A with the $l \times n$ matrix B is the $m \times n$ matrix C for which

$$c_{ij} = \sum_{k=1}^{l} a_{ik} b_{kj}, \tag{1.1.1}$$

and we write $C = AB$.

Notice that the operation (iii), known as matrix multiplication, is defined for *conformable* matrices. That is, the number of columns of the first factor must be equal to the number of rows of the second.

With these definitions, all three operations are associative and the first two are also commutative. Operation (ii) is distributive with respect to addition of elements of the field and addition of matrices, and (iii) is distributive with respect to addition of matrices.

We emphasize that in general $AB \neq BA$, but where the operation is commutative and $AB = BA$, we may say that A and B *commute*, or are permutable. If A and B commute, then they must obviously be square matrices of the same order.

From the definitions of matrix multiplication and of the transpose of a matrix it can easily be shown that

$$(AB)' = B'A',$$

and hence

$$(AB \ldots K)' = K' \ldots B'A'. \tag{1.1.2}$$

Thus, the transpose of a product of matrices is the product of their transposes taken in the reverse order.

If the $m \times n$ matrix A, $m \times l$ matrix B, $k \times n$ matrix C, and the $k \times l$ matrix D are written in the array

$$\begin{bmatrix} A & B \\ C & D \end{bmatrix}$$

they clearly form an $(m + k) \times (n + l)$ matrix, G, say. A, B, C, D may then be referred to as *partitions* of G and denoted by G_{11}, G_{12}, G_{21}, and G_{22} respectively. This idea can obviously be extended to include several partitions among the rows and columns of a matrix.

If matrices A and B are conformable and have partitions A_{ij} and B_{ij}, then it can be proved from the definition (iii) that the partitions of AB are given by

$$(AB)_{ij} = \sum_k A_{ik}B_{kj} \tag{1.1.3}$$

provided all the products appearing under the summation sign have a meaning, that is, provided the matrices of each pair A_{ik}, B_{kj} are conformable. Thus the product in terms of the partitions is obtained formally in the same way as that in terms of the scalar elements (cf. (1.1.1)). A detailed treatment of this topic is given by Aitken [1], §11.

Operations (i) and (ii) defined above are sufficient to ensure that all matrices of the same order with elements in F form a *linear space* over F (Gantmacher [1], Mirsky [1]). The most important of these spaces will be the *vector spaces* of all $n \times 1$ column matrices over the real and complex numbers. We denote these vector spaces by $\mathscr{R}_n$, $\mathscr{C}_n$ respectively.

1.2 Column and Row Vectors

An $n \times 1$ matrix (having only one column) is defined by its n scalar elements from the field F. Such matrices may be said to define sets of coordinates of points of an n-dimensional space. These coordinates define position vectors in the space, and this interpretation is emphasized by describing such an $n \times 1$ matrix as a *column vector* (or merely a vector if the context is clear) of order n. Such matrices or vectors will generally be distinguished by bold type. Thus $\boldsymbol{r}$ is a column vector whose elements will be denoted by $r_1, r_2, \ldots, r_n$.

Clearly the $1 \times n$ matrices may be interpreted in the same way, and these are called *row vectors*. They can always be written as the transpose of a column vector, and this will be our practice.

If $\boldsymbol{r}$ and $\boldsymbol{s}$ are vectors of order n (>1), the (matrix) products $\boldsymbol{r}'\boldsymbol{s}$ $(= \boldsymbol{s}'\boldsymbol{r})$ and $\boldsymbol{r}\boldsymbol{s}'$ are known as the *inner* and *outer products* of $\boldsymbol{r}$ and $\boldsymbol{s}$ respectively, and we note that the inner product is an element of F whereas the outer product is a square matrix of order n. The outer product, however, still has a meaning if $\boldsymbol{r}$ and $\mathbf{s}$ do not have the same order. If the inner product of two vectors is zero we say that they are *orthogonal.*

We may now write the set of equations

$$\begin{matrix} a_{11}x_1 + a_{12}x_2 + \cdots + a_{1n}x_n = y_1 \\ a_{21}x_1 + a_{22}x_2 + \cdots + a_{2n}x_n = y_2 \\ \vdots \qquad \vdots \qquad\qquad \vdots \quad \vdots \\ a_{m1}x_1 + a_{m2}x_2 + \cdots + a_{mn}x_n = y_m \end{matrix} \tag{1.2.1}$$

in the form

$$A\boldsymbol{x} = \boldsymbol{y}, \tag{1.2.2}$$

where $\boldsymbol{x}$ and $\boldsymbol{y}$ are vectors of order n and m respectively.

We will find the following result useful in subsequent matrix manipulations. It is an important special case of (1.1.3).

MULTIPLICATION LEMMA *Let A, B be conformable matrices and let $\boldsymbol{a}_1, \boldsymbol{a}_2, \ldots, \boldsymbol{a}_n$ be the columns of A and $\mathbf{b}'_1, \mathbf{b}'_2, \ldots, \mathbf{b}'_n$ be the rows of B, then*

$$AB = \sum_{k=1}^{n} \boldsymbol{a}_k \mathbf{b}'_k.$$

Proof Let A and B be of orders $m \times n$ and $n \times l$ respectively, and let $AB = C$, an $m \times l$ matrix. Then using (1.1.1) we have

$$c_{ij} = \sum_{k=1}^{n} a_{ik}b_{kj}, \qquad 1 \leqq i \leqq m, \quad 1 \leqq j \leqq l.$$

Now the outer product $\boldsymbol{a}_k\mathbf{b}'_k$ is an $m \times l$ matrix, and its (i, j) element is clearly $a_{ik}b_{kj}$. Hence the (i, j) element of $\sum \boldsymbol{a}_k\mathbf{b}'_k$ is just c_{ij}, and the lemma is proved.

1.3 SQUARE MATRICES

The *main diagonal* of a square matrix is made up of those elements a_{ij} for which $i = j$. A square matrix in which all elements off the main diagonal are zero is called a *diagonal matrix.* A *unit matrix* is a diagonal matrix for which all the main diagonal elements are ones, and is denoted by I, or possibility I_n if the order is not clear from the context.

If I is of order n, A is $m \times n$, and B is $n \times l$, then

$$AI = A \qquad \text{and} \qquad IB = B, \tag{1.3.1}$$

and

$$II = I^2 = I.$$

The determinant of a square matrix is that determinant whose elements are those of the matrix, and we write the determinant of A as $|A|$. We will use the product law:

$$|AB \ldots K| = |A| \,.\, |B| \ldots |K|. \tag{1.3.2}$$

That is, the determinant of a product of square matrices is the product of the determinants.

If $|A|$ is defined and is non-zero, then A is said to be *non-singular*, whereas A is *singular* if $|A| = 0$.

If the square matrix A has elements a_{ij} and if f_{ij} is the cofactor † of a_{ij}, the matrix with elements f_{ji} is called the *adjoint* of A. Thus, if F is the adjoint of A, we have for the (i, j) element of AF:

$$\sum_k a_{ik} f_{kj} = |A| \, \delta_{ij},$$

$F = ((-1)^{i+j} M_{ij})^T$

where δ_{ij} is the Kronecker delta. Thus we obtain

$$AF = |A| \, I, \tag{1.3.3}$$

and similarly,

$$FA = |A| \, I. \tag{1.3.4}$$

For a non-singular matrix $|A| \neq 0$, and we define the matrix $F/|A|$ as the *reciprocal* or *inverse* of A and write

$$A^{-1} = F/|A|. \tag{1.3.5}$$

This nomenclature is the natural outcome of (1.3.3) and (1.3.4) from which

$$A^{-1}A = AA^{-1} = I. \tag{1.3.6}$$

From these definitions it can be shown that

$$(AB \ldots K)^{-1} = K^{-1} \ldots B^{-1}A^{-1}. \tag{1.3.7}$$

Thus, the inverse of a product of non-singular square matrices is the product of the inverses taken in the reverse order.

† Let A be an $n \times n$ matrix. The minor α_{ij}, of order $n - 1$ of $|A|$ is the determinant of order $n - 1$ obtained from $|A|$ by striking out the ith row and jth column. The cofactor f_{ij} is given by $f_{ij} = (-1)^{i+j}\alpha_{ij}$.

If the matrix A of (1.2.2) is non-singular ($m = n$, and $|A| \neq 0$), then the unique solution of that equation is obtained by multiplying on the left by A^{-1} and using (1.3.6) and (1.3.1):

$$\boldsymbol{x} = A^{-1}\boldsymbol{y}.$$

We note also that the operations of taking the transpose and inverse of a non-singular matrix are permutable. Thus,

$$(A^{-1})' = (A')^{-1}. \tag{1.3.8}$$

There is clearly no ambiguity if we write $AA = A^2$, and the associativity of matrix multiplication implies that $A^2A = AA^2$, which we write as A^3. More generally we can define all positive integral powers of a square matrix and deduce that, for positive integers a and b,

$$A^aA^b = A^bA^a = A^{a+b}, \tag{1.3.9}$$

and

$$(A^a)^b = (A^b)^a = A^{ab}. \tag{1.3.10}$$

From the properties of the inverse of a *non-singular* matrix, (1.3.6) and (1.3.7), we can immediately extend the above index laws to include negative integral indices, provided we write $A^0 = I$.

Having defined the integral powers of a matrix we can define a polynomial of a matrix in the form

$$p(A) = a_0A^l + a_1A^{l-1} + \dots + a_{l-1}A + a_lI, \tag{1.3.11}$$

where $a_0, a_1, \dots, a_l$ are scalars.

We may also define a sequence of matrices such that the rth member, $A_r = \alpha_rA^r$, where the α_r are scalars. The associated power series $\sum_{r=0}^{\infty} A_r$ is defined to be convergent if every element of the partial sum $\sum_{r=0}^{m} A_r$ converges to a finite limit as $m \to \infty$. In particular we define the exponential function of a square matrix by

$$e^A = I + A + \frac{A^2}{2!} + \frac{A^3}{3!} + \dots, \tag{1.3.12}$$

which is easily shown to converge for all A (Frazer *et al.* [1], §2.5; Bellman [1], p. 166).

1.4 Linear Dependence, Rank, and Degeneracy

The vectors $\boldsymbol{q}_1, \boldsymbol{q}_2, \ldots, \boldsymbol{q}_n$ are *linearly dependent* if there exist constants $a_1, a_2, \ldots, a_n$ (from the field of the vectors $\boldsymbol{q}_i$) which are not all zero and are such that

$$a_1\boldsymbol{q}_1 + a_2\boldsymbol{q}_2 + \cdots + a_n\boldsymbol{q}_n = \boldsymbol{0}. \tag{1.4.1}$$

The same vectors are *linearly independent* if (1.4.1) is satisfied by $a_1 = a_2 = \ldots = a_n = 0$ only.

An $m \times n$ matrix is said to have *rank* r if r is the order of the largest non-vanishing minor† of A. Clearly $r \leqq \min (m, n)$.

This definition can be proved to be equivalent to the following: The rank r is the largest number of linearly independent rows of A. And a third equivalent definition is obtained if we replace "rows" in the latter by "columns." We notice that a matrix formed by the outer product of two non-zero vectors ($\boldsymbol{rs}'$ of § 1.2) has rank one.

If a square matrix A of order n has rank n it is obviously non-singular and its inverse exists. If the rank is $n - \alpha$ $(0 < \alpha < n)$ then there are α independent linear relations between the columns of A, and we may say that A has *degeneracy*, or *nullity*, α.

Consider the set of homogeneous algebraic equations in n unknowns represented by

$$A\boldsymbol{x} = \boldsymbol{0}. \tag{1.4.2}$$

If A is non-singular the only solution of this equation is obviously $\boldsymbol{x} = \boldsymbol{0}$; and it can be shown that (1.4.2) has exactly α linearly independent solutions different from the zero vector if and only if A has degeneracy α.

The following results concerning the rank of matrix products and sums are important and are used as required without further reference:

(i) The rank of a product of two rectangular matrices cannot exceed the rank of either matrix.

(ii) If a rectangular matrix of rank r is multiplied on the left or on the right by a non-singular square matrix, then the rank of the product matrix is r.

† A minor of order p ($p \leqq \min (m, n)$) is obtained from A by taking the determinant of the $p \times p$ matrix formed by striking out any $m - p$ rows and any $n - p$ columns of A.

(iii) The rank of the sum of two matrices cannot exceed the sum of the ranks of the two matrices.

The vector space $\mathscr{R}_n$ (or $\mathscr{C}_n$) contains at most n linearly independent vectors, so that any set of $n + 1$ vectors is necessarily linearly dependent. A *subspace* of $\mathscr{R}_n$ (or $\mathscr{C}_n$) is a set, $\mathscr{L}$, of vectors belonging to $\mathscr{R}_n$ (or $\mathscr{C}_n$) such that, for any two members $\boldsymbol{x}$, $\boldsymbol{y}$ of $\mathscr{L}$ and any two numbers a, b of F, the linear combination $a\boldsymbol{x} + b\boldsymbol{y}$ is in $\mathscr{L}$. The *dimension* of a subspace is now defined as that number p such that there exists a set of p linearly independent vectors in the subspace, while all sets of $p + 1$ vectors are linearly dependent. Any set of p linearly independent vectors in a subspace of dimension p is called a *basis* for the subspace. The notions of dimension and basis obviously apply to both $\mathscr{R}_n$ and $\mathscr{C}_n$.

Now consider α linearly independent solutions of equation (1.4.2). It is clear that *any* linear combination of these solutions is itself a solution of (1.4.2), and the set of all such combinations constitutes a subspace of dimension α. This example is particularly important; the subspace defined is known as the *nullspace* of A.

A set of vectors is said to *span* the subspace $\mathscr{L}$ if every vector in $\mathscr{L}$ may be expressed as a linear combination of the given set. By definition, a set of basis vectors for $\mathscr{L}$ will span $\mathscr{L}$ and, if $\mathscr{L}$ has dimension p, then a set of vectors which span $\mathscr{L}$ must contain at least p members.

1.5 Special Kinds of Matrices

If a matrix A has elements a_{ij}, and if $\bar{a}_{ij}$ denotes the complex conjugate of a_{ij}, then the matrix $\bar{A}$ with elements $\bar{a}_{ij}$ is called the complex conjugate of A. We define the following special kinds of matrix:

If $A' = A$, we say that A is *symmetric.*
If $A' = -A$, we say that is A *skew-symmetric.*
If $A'A = I$, we say that A is *orthogonal.*
If $\bar{A}' = A$, we say that A is *Hermitian.*
If $\bar{A}'A = I$, we say that A is *unitary.*

We pause here to note that any square matrix can be expressed uniquely as the sum of a symmetric matrix A_s and a skew-symmetric matrix A_k. Thus, if $a_{ij}^{(s)}$, a_{ij} and $a_{ij}^{(k)}$ are the elements of A, A_s, and A_k respectively, we have

$$a_{ij} = a_{ij}^{(s)} + a_{ij}^{(k)},$$

where

$$a_{ij}^{(s)} = \tfrac{1}{2}(a_{ij} + a_{ji}), \qquad a_{ij}^{(k)} = \tfrac{1}{2}(a_{ij} - a_{ji}).$$

Or, in matrix form,

$$A = A_s + A_k, \tag{1.5.1}$$

where

$$A_s = \tfrac{1}{2}(A + A'), \qquad A_k = \tfrac{1}{2}(A - A'). \tag{1.5.2}$$

For any matrix A an expression of the form $\boldsymbol{r}'A\boldsymbol{q}$ is a scalar, and is known as a *bilinear form* in the elements of $\boldsymbol{q}$ and $\boldsymbol{r}$. If A is a square matrix the expression $\boldsymbol{q}'A\boldsymbol{q}$ is a *quadratic form* in the elements of $\boldsymbol{q}$, and may also be written:

$$\boldsymbol{q}'A\boldsymbol{q} = \sum_j \sum_i a_{ij}q_iq_j. \tag{1.5.3}$$

We observe that, if A is skew-symmetric, equal to A_k, say, then

$$\boldsymbol{q}'A_k\boldsymbol{q} = 0. \tag{1.5.4}$$

We will describe the *real* square matrix A as *positive definite* if

$$\boldsymbol{q}'A\boldsymbol{q} > 0$$

for all real vectors $\boldsymbol{q}$ other than the zero vector. From (1.5.1) and (1.5.4) we deduce that, if A is positive definite, then so is its symmetric part A_s, and conversely. An obvious extension leads to the definition of a real matrix A as *non-negative* definite if

$$\boldsymbol{q}'A\boldsymbol{q} \geqq 0$$

for all $\boldsymbol{q}$ and if the equality sign holds for some $\boldsymbol{q} \neq \boldsymbol{0}$.

Suppose now that $\boldsymbol{q}$ is a vector with elements from the field of complex numbers, and A is again a real square matrix. Consider the complex scalar $\bar{\boldsymbol{q}}'A\boldsymbol{q}$. In an obvious notation we write $\boldsymbol{q} = \boldsymbol{q}_R + i\boldsymbol{q}_I$, and using the decomposition (1.5.1) we have

$$\begin{aligned} \bar{\boldsymbol{q}}'A\boldsymbol{q} &= (\boldsymbol{q}_R' - i\boldsymbol{q}_I')(A_s + A_k)(\boldsymbol{q}_R + i\boldsymbol{q}_I) \\ &= \boldsymbol{q}_R' A_s\boldsymbol{q}_R + \boldsymbol{q}_I'A_s\boldsymbol{q}_I + 2i\boldsymbol{q}_RA_k\boldsymbol{q}_I, \end{aligned} \tag{1.5.5}$$

using (1.5.4) and the fact that $\boldsymbol{q}_R' A_s\boldsymbol{q}_I = \boldsymbol{q}_I'A_s\boldsymbol{q}_R$, and $\boldsymbol{q}_I'A_k\boldsymbol{q}_R = -\boldsymbol{q}_R'A_k\boldsymbol{q}_I$. From (1.5.5) we deduce the following results:

(i) If A is a real symmetric positive definite matrix, then

$$\bar{\boldsymbol{q}}' A \boldsymbol{q} > 0 \tag{1.5.6}$$

for all complex vectors $\boldsymbol{q}$ other than the zero vector.

(ii) If A is a real skew-symmetric matrix, then

$$\text{Real part of } (\bar{\boldsymbol{q}}' A \boldsymbol{q}) = 0 \tag{1.5.7}$$

for all complex vectors $\boldsymbol{q}$.

1.6 Matrices Dependent on a Scalar Parameter; Latent Roots and Vectors

We consider a square matrix A in which the elements are defined as functions of a scalar parameter, λ. In defining integrals and differential coefficients of such matrices we will always assume the elements to be integrable and differentiable for as many times as are required by the current discussion. The derivative of $A(\lambda)$ with respect to λ is defined as that matrix whose elements are the first derivatives of the elements of A. We deduce immediately that

$$\frac{d}{d\lambda}\left\{A(\lambda) + B(\lambda)\right\} = \frac{dA}{d\lambda} + \frac{dB}{d\lambda}$$

and

$$\frac{d}{d\lambda}(AB) = \frac{dA}{d\lambda}B + A\frac{dB}{d\lambda},$$

where it is important to note that the order of the factors must always be preserved. This implies that in general

$$\frac{d}{d\lambda}(A^n) \neq nA^{n-1}\frac{dA}{d\lambda}$$

for integers $n > 1$, as one might hope, but that

$$\frac{d}{d\lambda}(A^n) = \sum_{i=1}^{n} A^{i-1}\frac{dA}{d\lambda}A^{n-i}.$$

If A is non-singular we have

$$A^{-n}A^n = I,$$

and, if A depends on λ,

$$\frac{dA^{-n}}{d\lambda}A^n + A^{-n}\frac{dA^n}{d\lambda} = 0.$$

Thus, the derivative of a negative integral power of A is given by

$$\frac{dA^{-n}}{d\lambda} = -A^{-n}\frac{dA^n}{d\lambda}A^{-n}. \tag{1.6.1}$$

We will also need to differentiate the exponential function with argument λA where A is a constant square matrix. From (1.3.12) we have

$$e^{\lambda A} = I + \lambda A + \frac{\lambda^2 A^2}{2!} + \frac{\lambda^3 A^3}{3!} + \cdots.$$

It can be shown that

$$\frac{d^n}{d\lambda^n}(e^{\lambda A}) = A^n e^{\lambda A} = e^{\lambda A}A^n, \tag{1.6.2}$$

for positive integers n.

The integral (definite or indefinite) of $A(\lambda)$ with respect to λ is simply defined as the matrix whose elements are the corresponding integrals of the elements of $A(\lambda)$.

In general, the determinant $|A(\lambda)|$ depends on λ and we define those values of λ for which $|A(\lambda)| = 0$ as the *latent roots* of $A(\lambda)$. If λ_i is any such root, then the sets of homogeneous equations

$$A(\lambda_i)\,\boldsymbol{q} = \boldsymbol{0} \qquad \text{and} \qquad \boldsymbol{r}'A(\lambda_i) = \boldsymbol{0}' \tag{1.6.3}$$

have at least one non-trivial solution for $\boldsymbol{q}$ and $\boldsymbol{r}$ respectively. The number of linearly independent solutions of either set is equal to the degeneracy of $A(\lambda_i)$ (cf. § 1.4). Any non-trivial solutions of (1.6.3) are known as *right* or *left latent vectors* of A, respectively. If $A(\lambda_i)$ has degeneracy α_i, then there are α_i linearly independent right latent vectors associated with the latent root λ_i, and similarly for the left latent vectors. If $A(\lambda)$ is a symmetric matrix, then the subspaces of right and left latent vectors associated with a particular root will coincide.

We note that latent roots and vectors are only defined for matrices dependent on a parameter. Thus the phrase "the latent roots (or vectors) of a constant matrix" will have no meaning.

1.7 Eigenvalues and Vectors

If A is a square matrix and I is the unit matrix of the same order, the *eigenvalues* of A are the zeros of $|A - \mu I|$, (viewed as a function of the scalar μ).

The *right* and *left eigenvectors* of A associated with the eigenvalue μ_i are the non-trivial solutions of

$$(A - \mu_i I)\,\boldsymbol{x} = \boldsymbol{0} \quad \text{and} \quad \boldsymbol{y}'(A - \mu_i I) = \boldsymbol{0}' \tag{1.7.1}$$

respectively.

We emphasize that, if A is a *constant* matrix, then the *latent roots of* $(A - \mu I)$ and the *eigenvalues of* A are synonymous, and similarly for the latent vectors of $(A - \mu I)$ and the eigenvectors of A. However, if the matrix A is a function of a parameter, λ, say, then the two notions are distinct. In this case the eigenvalues μ_i are dependent on λ, as are the eigenvectors. The latent roots and vectors, on the other hand, do not depend on λ. We also observe that, at best, the latent vectors and the eigenvectors are defined only to within arbitrary scalar multipliers.

The expression $|A - \mu I|$ is easily seen to be a polynomial in μ of degree n, the order of A. If the *distinct* eigenvalues of A are $\mu_1, \mu_2, \ldots, \mu_s$, then there exist positive integers $m_1, m_2, \ldots, m_s$ such that

$$|A - \mu I| = (\mu_1 - \mu)^{m_1}(\mu_2 - \mu)^{m_2} \cdots (\mu_s - \mu)^{m_s},$$

and

$$m_1 + m_2 + \ldots + m_s = n.$$

The numbers $m_1, m_2, \ldots, m_s$ are called the *multiplicities* of the respective eigenvalues.

We now prove two theorems which lie behind the basic assumptions of most of the work appearing in the sequel.

THEOREM 1.1 *A constant matrix A of order n and degeneracy α has a null eigenvalue with multiplicity $m \geq \alpha$.*

Proof By definition the eigenvalues of A are the solutions of the *characteristic equation* of A,

$$c(\mu) = |A - \mu I| = 0. \tag{1.7.2}$$

On expanding this determinant let

$$c(\mu) = (-1)^n \mu^n + c_1\mu^{n-1} + \cdots + c_{n-1}\mu + c_n = 0.$$

Now by our first definition of rank (§ 1.4) *all* minors of A of order greater than $r = n - \alpha$ are zero, and the coefficient c_p of the above equation is clearly a linear combination of certain pth order minors of A. Thus we have

$$c_{n-\alpha+1} = \cdots = c_{n-1} = c_n = 0,$$

and

$$c(\mu) = \{(-1)^n \mu^{n-\alpha} + c_1\mu^{n-\alpha-1} + \cdots + c_{n-\alpha}\}\,\mu^\alpha = 0.$$

Thus there exists a null eigenvalue whose multiplicity is at least α.

To show that we may have $m > \alpha$ it is only necessary to study the example

$$H = \begin{bmatrix} 0 & 1 \\ 0 & 0 \end{bmatrix}.$$

It is clear that the rank of H is 1 so that $\alpha = n - r = 1$. However, H has a two-fold null eigenvalue. Thus $m = 2 > \alpha$.

We now generalize this theorem to the following:

THEOREM 1.2 *If μ_i is an eigenvalue of A and the matrix $A - \mu_i I$ has degeneracy α, then μ_i has multiplicity $m \geqq \alpha$.*

Proof By the previous theorem we see that $A - \mu_i I$ has a null eigenvalue of multiplicity $m \geqq \alpha$. Furthermore, for any number k we have

$$|(A - kI) - (\mu_i - k)\, I| = 0,$$

which implies that the eigenvalues of $A - kI$ are those of A decreased by k. In particular, the eigenvalues of A are those of $A - \mu_i I$ increased by μ_i, and since $A - \mu_i I$ has a null eigenvalue of multiplicity m, A must have the eigenvalue μ_i with the same multiplicity.

We can also state the theorem in the following way:

COROLLARY *If α is the dimension of the subspace of right (or left) eigenvectors associated with the eigenvalue μ_i of A, and if μ_i has multiplicity m, then $m \geqq \alpha$.*

We now return to a matrix which is a function of λ, written $A(\lambda)$. If λ_i is a latent root of $A(\lambda)$, then the right latent vectors of λ_i are the solutions of

$$A(\lambda_i)\, \boldsymbol{q} = \boldsymbol{0}.$$

However, the right eigenvectors associated with the null eigenvalue of the constant matrix $A(\lambda_i)$ are defined as the solutions of

$$A(\lambda_i)\, \boldsymbol{x} = \boldsymbol{0}.$$

Hence these two sets of vectors must coincide and we have:

THEOREM 1.3 *If λ_i is a latent root of $A(\lambda)$, then the subspace spanned by the right eigenvectors of the null eigenvalues of $A(\lambda_i)$ coincides with the subspace spanned by the right latent vectors of λ_i.*

Notice here that the dimension of the subspace referred to in this theorem (i.e., the maximum number of linearly independent vectors in the subspace) is determined by the degeneracy of $A(\lambda_i)$, say α. The multiplicities of the latent root or of the null eigenvalue are not relevant here, and either or both of these quantities may be greater than α (the latter following from Theorem 1.1). In Theorem 1.3 we may obviously replace the word "right" by "left" everywhere.

The Hermitian matrices (§ 1.5) are a particularly simple but important class of matrices. The real symmetric matrices are included in this class, and we mention just two of their properties, though they are easily extended to include all Hermitian matrices.

First, we observe that all the eigenvalues of a real symmetric matrix, A, are real; for, given an eigenvalue μ with eigenvector $\boldsymbol{x}$, we have $\mu\boldsymbol{x} = A\boldsymbol{x}$, whence $\mu = \overline{\boldsymbol{x}}'A\boldsymbol{x}$, provided $\overline{\boldsymbol{x}}'\boldsymbol{x} = 1$, which we may assume to be the case, since $\boldsymbol{x} \neq \boldsymbol{0}$. Comparison with equation (1.5.5) immediately implies that $\overline{\boldsymbol{x}}'A\boldsymbol{x}$ is real, so that every eigenvalue is real. Any eigenvector now belongs to the null-space of a real matrix $A - \mu I$, whence all the eigenvectors may also be assumed to be real.

The eigenvalues of a Hermitian matrix can be defined by a second independent approach called the variational method. This is described in detail by Gantmacher [1] and Gould [1], and provides some very useful and illuminating material. We will not have occasion to use the variational approach, but we will need the following result, whose proof we omit, but which is of great importance in the development of the variational method.

Let A be a real symmetric matrix of order n with eigenvalues

$$\mu_1 \leqq \mu_2 \leqq \cdots \leqq \mu_n,$$

then for any vector $\boldsymbol{x}$ of $\mathscr{C}_n$ for which $\overline{\boldsymbol{x}}'\boldsymbol{x} = 1$,

$$\mu_1 \leqq \overline{\boldsymbol{x}}'A\boldsymbol{x} \leqq \mu_n. \tag{1.7.3}$$

It is easily seen that equality holds on the left (on the right) when $\boldsymbol{x}$ is an eigenvector associated with μ_1 (with μ_n).

1.8 Equivalent Matrices and Similar Matrices

We say that the matrix A, with elements in a field F, is *equivalent* to the matrix B if there exist non-singular matrices P and Q with elements in F such that $B = PAQ$. We may also say that A and B are connected by an *equivalence relation*.† Equivalence

† The theory is usually developed by building up P and Q as products of *elementary matrices*. See Gantmacher [1], Chapter 3; Mirsky [1], Chapter 6, etc.

relations are obviously reflexive, symmetric, and transitive; and it follows that equivalent matrices (in a given field) form a group.

It is also obvious that, if A is non-singular, then so are all matrices which are equivalent to A. More generally, it can be proved that *all square matrices of order n and rank r are equivalent in the field of their elements*, and in particular, are equivalent to the partitioned matrix

$$\begin{bmatrix} I_r & 0_{r,\,n-r} \\ 0_{n-r,\,r} & 0_{n-r} \end{bmatrix}$$

where subscripts denote the dimensions of the partitions.

If there exists a non-singular matrix T such that

$$B = T^{-1}AT \tag{1.8.1}$$

we say that A and B are *similar*, or that they are connected by a *similarity transformation*.† All matrices which are similar to one another will obviously form a subgroup of a group of equivalent matrices.

The most important property common to similar matrices is the fact that they all have the same eigenvalues, for

$$|A - \mu I| = 0$$

implies

$$|T^{-1}(A - \mu I)\,T| = 0,$$

and

$$|(T^{-1}AT - \mu I)| = 0,$$

so that

$$|B - \mu I| = 0$$

if (1.8.1) is satisfied. Thus the zeros of $|A - \mu I|$ coincide with those of $|B - \mu I|$ and the eigenvalues of A and B therefore coincide.

The converse statement is not necessarily true, i.e., matrices having the same eigenvalues are not necessarily similar (Gantmacher [1], Chapter 6, Theorem 7). For example, each of the two matrices

$$\begin{bmatrix} 0 & 0 \\ 0 & 0 \end{bmatrix} \quad \text{and} \quad \begin{bmatrix} 0 & 1 \\ 0 & 0 \end{bmatrix}$$

has a double eigenvalue, $\mu = 0$, but they are not similar.

† Also known as a collineatory transformation.

If (1.8.1) is satisfied we also have

$$B^2 = (T^{-1}AT)(T^{-1}AT) = T^{-1}A^2T,$$

and more generally

$$B^m = T^{-1}A^mT, \tag{1.8.2}$$

for any positive integer m. Thus for the polynomial $p(B)$ as defined by (1.3.11) we have

$$p(B) = T^{-1}p(A)\,T. \tag{1.8.3}$$

That is, if A and B are connected by a similarity transformation, then the polynomial matrices $p(A)$ and $p(B)$ are connected by the same transformation.

Consider now the $n \times n$ matrix A together with its n eigenvalues μ_i $(1 \leqq i \leqq n)$, which are not necessarily distinct. For each distinct μ_i select a set of basis vectors for the subspace of right eigenvectors. Let $\boldsymbol{x}_1, \boldsymbol{x}_2, \ldots, \boldsymbol{x}_m$ contain each of these basis vectors for each distinct μ_i and only these. We deduce from Theorem 1.2 that $m \leqq n$. Now form the square matrix X whose columns are the vectors $\boldsymbol{x}_1, \boldsymbol{x}_2, \ldots, \boldsymbol{x}_m$ followed by $n - m$ columns of zeros, and form the *diagonal* matrix of eigenvalues, U, with the eigenvalues corresponding to $\boldsymbol{x}_1, \ldots, \boldsymbol{x}_m$ in the first m positions. Then the m equations

$$A\boldsymbol{x}_j = \mu_j\boldsymbol{x}_j, \qquad j = 1, 2, \ldots, m,$$

are all included in the single relation

$$AX = XU. \tag{1.8.4}$$

We will prove shortly (Theorem 2.9, which is proved in a rather more general context) that right eigenvectors corresponding to pairwise distinct eigenvalues are linearly independent. Or, in other words, the intersection of any two subspaces of right eigenvectors corresponding to distinct eigenvalues is empty. Thus, matrix X of (1.8.4) has rank m. We now say that matrix A has *simple structure* if $m = n$, that is, if X is non-singular and the set of eigenvectors $\boldsymbol{x}_1, \boldsymbol{x}_2, \ldots, \boldsymbol{x}_n$ forms a basis for the whole n-dimensional vector space (over the field of the eigenvalues). We may also say that A has simple structure if and only if the dimension of each subspace of right eigenvectors is equal to the multiplicity of the corresponding eigenvalue.

Simple Structure

A has simple structure ⇔ square

In this case we deduce from (1.8.4) that

$$A = XUX^{-1}, \tag{1.8.5}$$

so that A and U are similar. On the other hand, if there exists a matrix X such that (1.8.5) is satisfied, we may then deduce (1.8.4) and observe that the n columns of X may be interpreted as linearly independent right eigenvectors of A. Thus we find:

THEOREM 1.4 *A square matrix A is similar to the diagonal matrix of its eigenvalues if and only if A has simple structure.*

We will see later, in § 2.2, that the linear independence of the right eigenvectors implies that of the left eigenvectors. A matrix which does not have simple structure will be said to be *defective.*

Probably the most important class of matrices with simple structure is that of the real symmetric matrices. It can be proved that every such matrix is orthogonally similar to the diagonal matrix of its eigenvalues.† In the above notation (with A now real and symmetric) there always exists an *orthogonal* matrix T such that

$$A = TUT^{-1}, \qquad \text{or} \qquad AT = TU; \tag{1.8.6}$$

and in addition the eigenvalues and vectors are real, so that U and T are real.

If A is a *complex* symmetric matrix, then it is not necessarily similar to a diagonal matrix. However, if A has simple structure, then there exists an orthogonal (complex) matrix T such that eqns. (1.8.6) are satisfied. The diagonal matrix U will then also be complex.

Now, if A is a matrix with an eigenvalue μ and right eigenvector $\boldsymbol{x}$, then $A\boldsymbol{x} = \mu\boldsymbol{x}$, and premultiplying by $\boldsymbol{x}'$ we find that

$$\mu = \frac{\boldsymbol{x}'A\boldsymbol{x}}{\boldsymbol{x}'\boldsymbol{x}} \tag{1.8.7}$$

provided $\boldsymbol{x}'\boldsymbol{x} \neq \boldsymbol{0}$. If A is a real symmetric matrix all the eigenvalues and vectors are real, so that $\boldsymbol{x}'\boldsymbol{x} \neq 0$ and (1.8.7) is certainly valid. Furthermore, if A is real, symmetric, and positive definite (§ 1.5), then the numerator of (1.8.7) is always positive, and we have the result that all the eigenvalues of such matrices are real and positive. We can therefore define a unique diagonal matrix $U^{1/2}$

† Appropriate theorems can be found in Bellman [1], Chapter 4; Ferrar [1], Chapter 6; Mirsky [1], Chapter 10; or Turnbull and Aitken [1], Chapter 8.

in terms of the positive square roots of the diagonal elements of U. Next define

$$A^{1/2} = TU^{1/2}T^{-1}, \tag{1.8.8}$$

where T is the orthogonal matrix of (1.8.6). Obviously, $(A^{1/2})^2 = A$, and $A^{1/2}$ is also a real, symmetric, positive definite matrix. Clearly, the word "positive" in this paragraph could be replaced everywhere by "non-negative."

1.9 The Jordan Canonical Form

We have seen that any simple matrix is similar to the diagonal matrix of its eigenvalues and we may now ask: What is the simplest form to which a general square matrix can be reduced by means of similarity transformations? Whatever the answer may be, we expect it to reduce to a diagonal form for simple matrices.

We have mentioned that eigenvectors corresponding to distinct eigenvalues are linearly independent; this implies that a non-simple square matrix must have at least one multiple eigenvalue and, furthermore, we know that in this case the multiplicity of at lease one eigenvalue must exceed the maximal number of linearly independent right eigenvectors associated with it. Such matrices may therefore be viewed as having coincident eigenvectors as well as coincident eigenvalues. This situation is sometimes described as a "confluence" of eigenvectors.

Before going on to give the general result we first introduce a rather unusual matrix. Define the $n \times n$ matrix H_n by

$$H_n = \begin{bmatrix} 0 & 1 & 0 & \cdots & 0 \\ 0 & 0 & 1 & \cdots & 0 \\ \vdots & \vdots & & & \vdots \\ 0 & 0 & \cdots & 0 & 1 \\ 0 & 0 & \cdots & 0 & 0 \end{bmatrix},$$

that is, having ones in the *superdiagonal* and zeros elsewhere. In terms of the Kronecker delta the elements of H_n may be written: $h_{ij} = \delta_{i,j-1}$. H_n has the following unusual properties which may easily be verified:

(i) $(H_n)^2$ has elements $\delta_{i,j-2}$ and $(H_n)^m = 0$ for integers $m \geq n$.

(ii) H_n has only one eigenvalue, namely, zero with multiplicity n.

(iii) H_n is not a simple matrix. The right eigenspace of the null eigenvalue has dimension one.

The following theorem now describes the Jordan canonical form.

THEOREM 1.5 *For an arbitrary $n \times n$ matrix A there exists a non-singular matrix T such that*

$$T^{-1}AT = \begin{bmatrix} J_{n_1}(\lambda_1) & 0 & \dots & 0 \\ 0 & J_{n_2}(\lambda_2) & \dots & 0 \\ \vdots & \vdots & \ddots & \vdots \\ 0 & \dots & 0 & J_{n_k}(\lambda_k) \end{bmatrix}, \qquad (1.9.1)$$

where

$$n_1 + n_2 + n_3 + \dots + n_k = n,$$
$$J_{n_i}(\lambda_i) = -\lambda_i I_{n_i} + H_{n_i}$$

and $\lambda_1, \lambda_2, \dots, \lambda_k$ are the eigenvalues of A (not necessairly distinct).

The matrices $J_{n_i}(\lambda_i)$ are called *Jordan blocks*. For example, if $n_1 = 3$, then

$$J_{n_1}(\lambda_1) = \begin{bmatrix} \lambda_1 & 1 & 0 \\ 0 & \lambda_1 & 1 \\ 0 & 0 & \lambda_1 \end{bmatrix}.$$

Notice also that, if $n_i = 1$, then $J_{n_i} = \lambda_i$, and, if this is true for every i, then the Jordan canonical form (1.9.1) reduces to a diagonal matrix. It may easily be verified that the maximal number of linearly independent right eigenvectors associated with an eigenvalue is the number of Jordan blocks in which the eigenvalue appears (cf. property (iii) of H_n). This number is known as the *index* of the eigenvalue, and it can be seen that Theorem 1.2 may be obtained from Theorem 1.5 in the following form: the multiplicity of an eigenvalue is not less than its index. If the index of each eigenvalue of a matrix is equal to its multiplicity, then all the Jordan blocks are of order one and the matrix is simple.†

Proofs of Theorem 1.5 may be found in Chapter 4 of Ferrar [1], Chapter 6 of Gantmacher [1] and Chapter 6 of Turnbull and Aitken [1].

† There is a close connection between this discussion and that of the elementary divisors of an eigenvalue. See also § 3.3.

1.10 Bounds for Eigenvalues

There are many situations in which we can make use of estimates of the magnitudes of eigenvalues, or of their approximate positions in the Argand diagram, provided these estimates are easily obtained. One of the most useful results of this kind is known as Geršgorin's theorem (Todd [1], § 8.14).

Theorem 1.6 *If A is a complex $n \times n$ matrix and*

$$\varrho_i = \sum{}' |a_{ij}|, \tag{1.10.1}$$

where $\sum'$ denotes the sum from $j = 1$ to n, $j \neq i$, then every eigenvalue of A lies in at least one of the discs

$$|z - a_{ii}| \leqq \varrho_i, \qquad i = 1, 2, \ldots, n. \tag{1.10.2}$$

Furthermore, if a set of m of these discs has no point in common with the remaining $n - m$ discs, then these discs contain m eigenvalues, if counted according to their multiplicities.

An analogous result holds in which the radii of the discs are determined by the *column* sums $\sum' |a_{ji}|$ instead of the *row*sum s (1.10.1).

Proof Let μ be an eigenvalue of A with right eigenvector $\boldsymbol{x}$. Then $A\boldsymbol{x} = \mu\boldsymbol{x}$, or, if we write $\boldsymbol{x}' = [x_1, x_2, \ldots, x_n]$, then

$$\sum_{j=1}^{n} a_{ij}x_j = \mu x_i, \qquad i = 1, 2, \ldots, n.$$

Of the numbers $x_1, x_2, \ldots, x_n$, let x_k be that of greatest modulus; then

$$\begin{aligned} |\mu - a_{kk}|\,|x_k| &= \left|\sum{}' a_{kj}x_j\right|, \\ &\leqq \sum{}' |a_{kj}|\,|x_j|, \\ &\leqq |x_k| \sum{}' |a_{kj}|. \end{aligned}$$

Thus, since $|x_k| \neq 0$, we have $|\mu - a_{kk}| \leqq \sum' |a_{kj}|$ and the first result is proved.

We now restrict our attention to the case $m = 1$. The proof for the general statement is immediately obvious once this case is settled. Thus, we suppose the ith disc to be distinct from all the others. Consider the matrix $B(t)$, where t is a scalar parameter. The elements of $B(t)$ are identically those of A with the exception of the b_{ij} (of the ith row) for which $j \neq i$. The latter are defined by $b_{ij} = ta_{ij}$. When $t = 1$, then $B(t) = A$, and the matrix $B(0)$ has the ith row with a_{ii} as diagonal element and zeros elsewhere. Hence,

a_{ii} is an eigenvalue of $B(0)$. Now let t increase continuously from $t = 0$ to $t = 1$ and observe the behavior of the Geršgorin discs for $B(t)$. The radius of the ith disc is proportional to t and all the other discs are invariant. Furthermore, the eigenvalues of a matrix are continuous functions of the matrix elements (Ostrowski [3], appendix K), so that the eigenvalue starting in the ith disc (as a_{ii}) at $t = 0$ must remain in this disc while $0 \leqq t \leqq 1$, and no other eigenvalue may enter the ith disc. This completes the proof for $m = 1$.

We will make use of a (necessarily more cumbersome) refinement of this theorem due to Brauer [1]. We state the theorem here without proof, using the notation of Theorem 1.6.

THEOREM 1.7 *Every characteristic root of A lies in at least one of the* $\frac{1}{2}n(n-1)$ *Cassini ovals:*

$$|z - a_{ii}|\,|z - a_{jj}| \leqq \varrho_i\varrho_j, \qquad i \neq j, \tag{1.10.3}$$

As above, a separation property can be established. If

$$\sqrt{\varrho_i\varrho_j} < \tfrac{1}{2}|a_{ii} - a_{jj}|, \tag{1.10.4}$$

then the corresponding oval is of two distinct segments (Fig. 1.1),

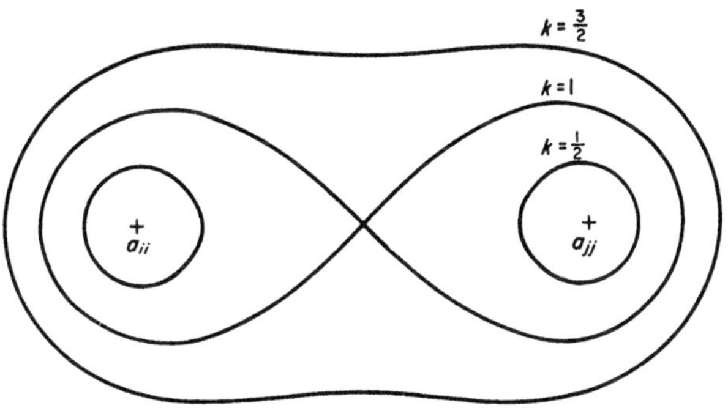

FIG. 1.1. Bounding curves for Brauer's theorem (Theorem 1.7).

one "centered" on a_{ii} and the other on a_{jj}. In Fig. 1.1 the parameter k is the quotient of $\sqrt{\varrho_i\varrho_j}$ and $\frac{1}{2}|a_{ii} - a_{jj}|$. Further, if (1.10.4) is satisfied for $j = 1, 2, \ldots, n$, $j \neq i$, and those segments of the $n - 1$ Cassini ovals containing a_{ii} have no point in common with the remaining $\frac{1}{2}(n-1)(n-2)$ ovals, then the union of these $n - 1$ regions contains one and only one eigenvalue of A.

Finally, we give a special case of a superposition theorem due to Bauer and Householder [1]. By a superposition theorem we mean, in this context, a result concerning the eigenvalues of the matrix sum $A + B$ given the eigenvalues of A and B. Our special case of Bauer and Householder's result may be formulated as follows:

THEOREM 1.8 *Let A and B be complex $n \times n$ matrices which are unitarily similar to diagonal matrices and let α be any scalar. If we write*

$$\varrho(\alpha, B) = \max_j |\mu_j(B) - \alpha|, \tag{1.10.5}$$

then each eigenvalue of $A + B$ lies in the union of the circular regions

$$|z - (\alpha + \mu_i(A))| \leq \varrho(\alpha, B), \qquad i = 1, 2, \ldots, n. \tag{1.10.6}$$

Let us first make it clear that matrices A and U are unitarily similar if there exists a unitary matrix T (§ 1.5) such that $A = T^{-1} UT$. In particular, a Hermitian matrix (which may be a real symmetric matrix) is unitarily similar to the diagonal matrix of its eigenvalues.

$\mu_i(A)$, $\mu_j(B)$ are, ofcourse, typical eigenvalues of A, B respectively. In order to see the significance of this result suppose first of all that $\alpha = 0$. Then $\varrho = \max_j |\mu_j(B)|$ and the circles (1.10.6) have radius ϱ and centers $\mu_i(A)$, $i = 1, 2, \ldots, n$. However, with a little reflection, it is easily seen that in general the radius of these circles can be reduced by some other choice of α; and the smaller ϱ is, the closer are the bounds which the theorem provides. Thus, we generally choose α to minimize the distance $\max_j |\mu_j(B) - \alpha|$.

pencil $A\lambda + C$

elements $\in$ *Complex*
$\{$
λ

Regular Pencil: A & C *square*
A *non-singular*

CHAPTER 2

REGULAR PENCILS OF MATRICES AND EIGENVALUE PROBLEMS

2.1 INTRODUCTION

If A and C are constant matrices of the same dimensions with elements in a field F, then all matrices of the form $A\lambda + C$, where λ is in F, constitute a pencil of matrices. In the sequel the field F will always be that of the complex numbers unless stated otherwise.

A pencil of matrices will be called *regular*† if (a) A and C are square matrices and (b) A is non-singular.

A regular pencil of matrices may also be described as a square matrix whose elements are linear functions of a parameter, λ, and in which the matrix coefficient of λ is non-singular. From the discussion of latent roots and vectors in § 1.6 we observe, therefore, that there is a set of n latent roots associated with a regular pencil of matrices of order n. These are the roots of the equation

$$|A\lambda + C| = 0, \tag{2.1.1}$$

and are n in number because the coefficient of λ^n in the expansion of this determinant is $|A|$, which is non-zero for a regular pencil. The multiplicity of a latent root λ_i is defined as the number of times the factor $(\lambda - \lambda_i)$ appears in the factorization of the polynomial $|A\lambda + C|$ into linear factors.

Clearly the discussion of eigenvalues and vectors in § 1.7 and § 1.8, where we considered the matrix $A - \mu I$, is related to a special kind of regular pencil. Thus any results developed here for a regular matrix pencil will also provide results for the eigenvalue problem as a special case.

---

† This definition does not coincide with that of Gantmacher [2], but is consistent with his definition of regular λ-matrices (Gantmacher [1], p. 76). See also § 3.1 of this book.

We now generalize (1.8.4) by first defining the $n \times n$ matrix

$$Q = [\boldsymbol{q}_1, \boldsymbol{q}_2, \ldots, \boldsymbol{q}_n], \tag{2.1.2}$$

where the $\boldsymbol{q}_i$ $(1 \leqq i \leqq m)$ are linearly independent right latent vectors of the regular matrix pencil $A\lambda + C$, and Q is filled up with columns of zeros if necessary (cf. (1.8.4)). Λ is then the corresponding diagonal matrix of latent roots, and we have

$$AQ\Lambda + CQ = 0. \tag{2.1.3}$$

Similarly, from the left latent vectors, $\boldsymbol{r}_i$, we form the $n \times n$ matrix

$$R = [\boldsymbol{r}_1, \boldsymbol{r}_2, \ldots, \boldsymbol{r}_n], \tag{2.1.4}$$

and we obtain the equation

$$\Lambda R'A + R'C = 0. \tag{2.1.5}$$

We note that, since the vectors $\boldsymbol{q}_i$, $\boldsymbol{r}_i$ are not defined to within an arbitrary scalar multiplier, the matrices Q and R are, at best, not defined to within an arbitrary non-singular diagonal matrix applied as a postfactor.

In the subsequent work we will consider a yet more restrictive class of pencils as follows. *A simple matrix pencil* of order n has the properties:

(a) It is a regular pencil.

(b) It has n linearly independent right latent vectors.

If the latter condition is not satisfied, the pencil will be said to be *defective.*

In this chapter we first establish some important theorems for simple matrix pencils. These results will be required later in the analysis of matrix polynomials. In § 2.4–2.8 we investigate the eigenvalue problem for matrices of simple structure and the properties of simple pencils of matrices. Finally we introduce two further (disconnected) ideas which we will also need for future reference.

2.2 Orthogonality Properties of the Latent Vectors

THEOREM 2.1 *If $A\lambda + C$ is a simple matrix pencil, then the right and left latent vectors of the pencil can be defined in such a way that*

$$R'AQ = I \quad \textit{and} \quad R'CQ = -\Lambda. \tag{2.2.1}$$

Proof By the definition of a simple matrix pencil the matrix A is non-singular so that (2.1.3) may be written

$$Q\Lambda + A^{-1}CQ = 0;$$

and since Q may also be chosen to be non-singular,

$$\Lambda Q^{-1} + Q^{-1}A^{-1}C = 0.$$

This may be written in the form

$$\Lambda(AQ)^{-1}A + (AQ)^{-1}C = 0,$$

and putting

$$R' = (AQ)^{-1} \tag{2.2.2}$$

we have

$$\Lambda R'A + R'C = 0.$$

Comparing this equation with (2.1.5) we observe that the matrix R of (2.2.2) may be interpreted as the matrix of left latent vectors, (2.1.4).

The first of equations (2.2.1) now follows from (2.2.2), and the second is obtained by premultiplying (2.1.3) by R'.

COROLLARY *Taking* (2.2.2) *as our definition for R we deduce that a simple matrix pencil of order n has n linearly independent left latent vectors.*

When Q and R satisfy the first of equations (2.2.1), the latent vectors are said to satisfy a biorthogonality condition with respect to the matrix A. Notice also that the conditions (2.2.1) may be written

$$\mathbf{r}_i'A\mathbf{q}_j = \delta_{ij} \quad \text{and} \quad \mathbf{r}_i'C\mathbf{q}_j = -\delta_{ij}\lambda_i, \tag{2.2.3}$$

where δ_{ij} is the Kronecker delta. The choice of the latent vectors (corresponding to multiple roots) which gives rise to this biorthogonality property warrants further examination.

LEMMA *A simple matrix pencil has α linearly independent right (or left) latent vectors associated with a latent root of multiplicity α.*

This lemma follows from the definition of a simple matrix pencil (for the right latent vectors) and from the above corollary (for the left latent vectors).

THEOREM 2.2 *Let the subspaces of the right and left latent vectors of a latent root of the matrix pencil $A\lambda + C$ each have dimension α and let the columns of $n \times \alpha$ matrices Q_α and R_α form bases for these subspaces satisfying*

$$R_\alpha'AQ_\alpha = I_\alpha; \tag{2.2.4}$$

then the matrices $Q_\alpha G$, $R_\alpha(G^{-1})'$ *have the same property, where* G *is an arbitrary non-singular matrix of order* α.

Proof We note first of all that Theorem 2.1 ensures that it is always possible to choose Q_α and R_α so that (2.2.4) is satisfied. Let $\mathcal{Q}_\alpha = Q_\alpha G$ and $\mathcal{R}_\alpha = R_\alpha(G^{-1})'$, then

$$\begin{aligned}\mathcal{R}'_\alpha A \mathcal{Q}_\alpha &= G^{-1}R'_\alpha A Q_\alpha G \\ &= G^{-1}G = I_\alpha,\end{aligned}$$

for an arbitrary non-singular matrix G of order α. This proves the theorem. We note that if A is symmetric the formal coincidence of right and left eigenvectors is maintained if we choose an orthogonal matrix for G, in which case $(G^{-1})' = G$.

In numerical analysis a "deflation" process is often useful. That is, given a pencil $A\lambda + C$ and a latent root λ_1, we may wish to find a second pencil having all the latent roots of $A\lambda + C$ with the exception of λ_1. The following result is of this kind.

THEOREM 2.3 *If the simple pencil* $A\lambda + C$ *has a non-zero latent root* λ_1, *with right and left latent vectors* $\boldsymbol{q}_1$, $\boldsymbol{r}_1$, *for which* $\boldsymbol{r}'_1 A \boldsymbol{q}_1 = 1$, *and if we define* $\boldsymbol{s}_1 = A\boldsymbol{q}_1$, $\boldsymbol{t}_1 = A'\boldsymbol{r}_1$, *then the pencil*

$$A\lambda + C + \lambda_1 \boldsymbol{s}_1 \boldsymbol{t}'_1 \tag{2.2.5}$$

is a simple pencil and has the latent roots of $A\lambda + C$ *with the exception that one of the roots equal to* λ_1 *is replaced by a null latent root.*

Proof For any λ, $\boldsymbol{q}$ we have

$$(A\lambda + C + \lambda_1 \boldsymbol{s}_1 \boldsymbol{t}'_1)\,\boldsymbol{q} = (A\lambda + C)\,\boldsymbol{q} + \lambda_1 A \boldsymbol{q}_1 (\boldsymbol{r}'_1 A \boldsymbol{q}). \tag{2.2.6}$$

Now if λ is a latent root of $A\lambda + C$ other than λ_1 with $\boldsymbol{q}$ as a right latent vector, then $(A\lambda + C)\,\boldsymbol{q} = \boldsymbol{0}$ by definition and $\boldsymbol{r}'_1 A \boldsymbol{q} = 0$ by (2.2.3). Hence, we deduce from the above equation that λ is also a latent root of (2.2.5). If $\lambda = \lambda_1$ and $\boldsymbol{q} = \boldsymbol{q}_1$, then we find that

$$(C + \lambda_1 \boldsymbol{s}_1 \boldsymbol{t}'_1)\,\boldsymbol{q}_1 = \boldsymbol{0},$$

which implies that $\boldsymbol{q}_1$ is a latent vector of (2.2.5) associated with a null latent root.

Suppose now that λ_1 has multiplicity $\alpha > 1$. Then there exist vectors $\boldsymbol{q}_2, \boldsymbol{q}_3, \ldots, \boldsymbol{q}_\alpha$ such that

$$\left.\begin{aligned}(A\lambda_1 + C)\,\boldsymbol{q}_i &= 0,\\ \boldsymbol{r}'_1 A \boldsymbol{q}_i &= 0\end{aligned}\right\}, \qquad i = 2, 3, \ldots, \alpha. \tag{2.2.7}$$

Using Theorem 2.1 and the lemma to 2.2 we see that this is always the case provided $A\lambda + C$ is a simple pencil. Putting $\lambda = \lambda_1$ and $\boldsymbol{q} = \boldsymbol{q}_i$ in (2.2.6) and using (2.2.7), we now deduce that the pencil (2.2.5) has λ_1 as a latent root with multiplicity $\alpha - 1$. That the pencil is simple follows from the fact that the latent vectors of (2.2.5) are just those of $A\lambda + C$. The theorem is now proved.

Clearly, if λ_1 is a multiple latent root of $A\lambda + C$, then we must calculate biorthogonal bases for the associated subspaces of right and left latent vectors before the effect of λ_1 can be removed altogether.

2.3 The Inverse of a Simple Matrix Pencil

Theorem 2.4 *If $A\lambda + C$ is a simple matrix pencil, then provided λ is not equal to a latent root of the pencil, Q and R can be defined in such a way that*

$$(A\lambda + C)^{-1} = Q(I\lambda - \Lambda)^{-1} R'. \tag{2.3.1}$$

Proof From equations (2.2.1) we deduce that Q and R may be defined so that

$$R'(A\lambda + C)\,Q = I\lambda - \Lambda.$$

If λ is not equal to a latent root we may invert both sides of this equation to obtain

$$Q^{-1}(A\lambda + C)^{-1}\,(R')^{-1} = (I\lambda - \Lambda)^{-1},$$

which implies (2.3.1) and proves the theorem.

The right-hand side of (2.3.1) can be written in another illuminating way. We recall the definitions of Q and R and note that the diagonal elements of $(I\lambda - \Lambda)^{-1}$ are $(\lambda - \lambda_1)^{-1}, \ldots, (\lambda - \lambda_n)^{-1}$; then the multiplication lemma of § 1.2 gives

$$(A\lambda + C)^{-1} = \sum_{i=1}^{n} \frac{F_i}{\lambda - \lambda_i}, \tag{2.3.2}$$

where

$$F_i = \boldsymbol{q}_i \boldsymbol{r}_i', \qquad i = 1, 2, \ldots, n. \tag{2.3.3}$$

That is, F_i are square matrices of order n and rank one. Using equations (2.2.3) it is easily verified that these matrices satisfy the relations

$$F_i A F_j = \delta_{ij} F_i \tag{2.3.4}$$

and

$$F_i C F_j = -\lambda_i \delta_{ij} F_i. \tag{2.3.5}$$

If the pencil of matrices is symmetric, that is, if A and C are symmetric, then the right and left latent vectors coincide and we write $\boldsymbol{r}_i = \boldsymbol{q}_i$ and $R = Q$. This gives rise to some simplification of our results (2.2.1), (2.2.3), (2.3.1), and (2.3.2).

2.4 Application to the Eigenvalue Problem

As we have mentioned earlier we may specialize the results obtained to give similar results for eigenvalues and vectors. Thus Theorems 2.1 and 2.4 give the following results for a matrix of simple structure (§ 1.8), the notation being that of § 1.7.

(i) The right and left eigenvectors of a matrix of simple structure can be defined in such a way that

$$Y'X = I \qquad \text{and} \qquad Y'AX = U, \tag{2.4.1}$$

or equivalently,

$$\boldsymbol{y}_i'\boldsymbol{x}_j = \delta_{ij} \qquad \text{and} \qquad \boldsymbol{y}_i'A\boldsymbol{x}_j = \delta_{ij}\mu_i. \tag{2.4.2}$$

We say that the systems of vectors $\boldsymbol{y}_i, \boldsymbol{x}_i$ $(i = 1, 2, \ldots, n)$ are biorthogonal.

We note that, if $\mathscr{X}_1, \mathscr{Y}_1$ denote the subspaces of right and left eigenvectors of an eigenvalue μ and if A has simple structure, then there exist bases $\boldsymbol{x}_1, \boldsymbol{x}_2, \ldots, \boldsymbol{x}_\alpha$ and $\boldsymbol{y}_1, \boldsymbol{y}_2, \ldots, \boldsymbol{y}_\alpha$ for $\mathscr{X}_1$, $\mathscr{Y}_1$ respectively such that

$$\boldsymbol{y}_i'\boldsymbol{x}_j = \delta_{ij}, \qquad i, j = 1, 2, \ldots, \alpha.$$

(ii) If A is a matrix of simple structure, and equations (2.4.1) are satisfied, then

$$(A - \mu I)^{-1} = X(U - \mu I)^{-1}\,Y' \tag{2.4.3}$$

provided μ is not an eigenvalue of A. Or,

$$(A - \mu I)^{-1} = \sum_{i=1}^{n} \frac{G_i}{\mu_i - \mu}, \tag{2.4.4}$$

where

$$G_i = \boldsymbol{x}_i\boldsymbol{y}_i', \qquad i = 1, 2, \ldots, n. \tag{2.4.5}$$

The matrix $(A - \mu I)^{-1}$ is known as the *resolvent* of A.

The matrices G_i satisfy the following conditions:

$$G_iG_j = 0 \qquad \text{if} \qquad i \neq j, \tag{2.4.6}$$

and

$$G_i^2 = G_i. \tag{2.4.7}$$

Matrices satisfying the latter condition are known as *idempotent* matrices, and the matrices G_i will be referred to as *constituent matrices* of A. They are also known as the *projectors* of A.

From (2.4.1) we also have

$$A = (Y')^{-1} U X^{-1},$$

with $X^{-1} = Y'$ and $(Y')^{-1} = X$, so that

$$A = XUY'.$$

Furthermore, we have

$$A^2 = (XUY')(XUY') = XU^2Y',$$

using the first of (2.4.1); and more generally, *if A is a matrix of simple structure, then for all non-negative integers, m:*

$$A^m = XU^mY' = \sum_{i=1}^{n} \mu^m G_i. \tag{2.4.8}$$

From (2.4.3) and the first of (2.4.1) we obtain similarly

$$(A - \mu I)^{-m} = X(U - \mu I)^{-m} Y', \quad (\mu \neq \mu_i).$$

Now, if A is non-singular, zero is not an eigenvalue of A and we can put $\mu = 0$ in this equation to obtain the following result: *If A is a non-singular matrix of simple structure, then for all integers m (positive, zero, or negative) we have*

$$A^m = XU^mY' = \sum_{i=1}^{n} \mu^m G_i. \tag{2.4.9}$$

It follows from (2.4.8) that for a scalar polynomial with a matrix argument, as defined in (1.3.11), we have

$$\varrho(A) = \sum_{i=1}^{n} \varrho(\mu_i)\, G_i, \tag{2.4.10}$$

provided A is of simple structure. This is a special case of Sylvester's theorem (cf. Frazer *et al.* [1], § 3.9, 3.10), which Gantmacher calls the *fundamental formula* for $\varrho(A)$ (p. 104 of Gantmacher [1]).

We immediately obtain from (2.4.10) a special case of the famous Cayley–Hamilton theorem, namely, that every matrix satisfies its own characteristic equation. However, we have proved the theorem here for matrices of simple structure only. A more general proof will occur later (§ 3.5).

Let us return to matrices whose elements are functions of the parameter λ. We note first of all that, because the eigenvalues are continuous functions of the matrix elements, the continuity of the matrix elements as functions of λ implies that the eigenvalues are also continuous functions of λ (Ostrowski [3], appendix K).

We now assume that the elements of $A(\lambda)$ are regular functions of λ in some neighborhood of λ_0 and spend the remainder of this section considering the implications for the eigenvalues of $A(\lambda)$ at and near the point λ_0 of the complex λ-plane. We note first of all that the characteristic equation $|A(\lambda) - \mu I| = 0$ is a polynomial of degree n in μ whose coefficients are regular functions of λ in some neighborhood of μ_0. The solutions $\mu(\lambda)$ of this equation are *algebraic functions*, and there is a classical result (Goursat [1]) to the effect that, if $\mu(\lambda_0)$ is a simple eigenvalue, then $\mu(\lambda)$ is a regular function in some neighborhood of λ_0. This leaves the nature of multiple eigenvalues at λ_0 to be investigated. In the case of a Hermitian matrix $A(\lambda)$, this question has been answered by Rellich [1], who shows that the eigenvalues $\mu(\lambda)$ which coincide at λ_0 are also (real) regular functions of λ. Unfortunately, this is not of much help for our purposes and we will consider theorems concerning the behavior of multiple eigenvalues of more general matrices. The proofs (Lancaster [5]) are not included as they are rather long and not relevant to our subsequent work.

In the sequel we denote the subspaces of right and left eigenvectors of an eigenvalue $\mu(\lambda_0)$ by $\mathscr{X}_1$, $\mathscr{Y}_1$ respectively, and we use indices for derivatives with respect to λ. For example, $\mu^{(2)}$ will denote $d^2\mu/d\lambda^2$.

THEOREM 2.5 *Let $A(\lambda_0)$ be a matrix of simple structure having an eigenvalue $\mu(\lambda_0)$ of multiplicity α. Let the columns of the $n \times \alpha$ matrices X_α, Y_α form biorthogonal systems which span $\mathscr{X}_1$, $\mathscr{Y}_1$ respectively; then if $A^{(q)}(\lambda_0)$ is the first non-vanishing derivative of $A(\lambda)$ at λ_0 we have:*

(i) *The eigenvalues of $A(\lambda)$ are q times differentiable*† *at λ_0 and their first $q - 1$ derivatives vanish at λ_0.*

(ii) *The* qth *derivatives of the eigenvalues at λ_0 for which $\mu(\lambda) \to \mu(\lambda_0)$ as $\lambda \to \lambda_0$ are given by the eigenvalues of the $\alpha \times \alpha$ matrix $Y'_\alpha A^{(q)}(\lambda_0) X_\alpha$.*

(iii) *The left and right eigenvectors of $A(\lambda)$ may be assumed to be differentiable at λ_0.*

Though we do not attempt to prove this theorem we can make

† But not necessarily continuously differentiable.

part (ii) plausible by the following argument. Let $\boldsymbol{x}(\lambda)$, $\boldsymbol{y}(\lambda)$ be eigenvectors of $\mu(\lambda)$ for which $\boldsymbol{y}'(\lambda)\,\boldsymbol{x}(\lambda) = 1$, and

$$(A - \mu I)\,\boldsymbol{x} = \boldsymbol{0}, \qquad \boldsymbol{y}'(A - \mu I) = \boldsymbol{0}'. \tag{2.4.11}$$

Differentiate the first of these equations with respect to λ, *assuming* μ and $\boldsymbol{x}$ to be differentiable, and we have

$$(A^{(1)} - \mu^{(1)}I)\,\boldsymbol{x} + (A - \mu I)\,\boldsymbol{x}^{(1)} = \boldsymbol{0}.$$

Now premultiply this equation by $\boldsymbol{y}'$ and, using the second of equations (2.4.11), we find that $\mu^{(1)} = \boldsymbol{y}'A^{(1)}\,\boldsymbol{x}$. This is a special case of result (ii).

It will generally be the case that $A^{(1)}(\lambda_0) \neq 0$ so that $q = 1$ and part (ii) of Theorem 2.5 will be used to calculate the first derivatives of the eigenvalues (see the remark following Theorem 4.5, in particular). The next two theorems could then be used to investigate the second derivatives.

We should also note that, for simplicity, the hypothesis of the theorem has been made stronger than necessary. For part (ii) of the theorem the matrix $A(\lambda_0)$ need not be of simple structure. The essential feature contained in the hypothesis is the property that the multiplicity of the eigenvalue $\mu(\lambda_0)$ should be equal to its index (cf. § 1.9). This will also apply to Theorem 2.7.

To illustrate Theorem 2.5 consider

$$A(\lambda) = \begin{bmatrix} 0 & \lambda - 1 \\ (\lambda - 1)\lambda & 0 \end{bmatrix}.$$

This matrix has simple structure if and only if $\lambda \neq 0$. The characteristic equation of A is

$$c(\mu) = \mu^2 - \lambda(\lambda - 1)^2 = 0$$

so that, for any λ, the eigenvalues of A are given by

$$\mu(\lambda) = \pm\lambda^{1/2}(\lambda - 1),$$

and we see that

$$\mu^{(1)}(\lambda) = \pm\tfrac{1}{2}\lambda^{-1/2}(3\lambda - 1).$$

Thus the eigenvalues are differentiable for all λ except $\lambda = 0$, which is the only value of λ for which $A(\lambda)$ is not of simple structure. The eigenvalues are differentiable at $\lambda = 1$, even though they coincide at this point.

THEOREM 2.6 *If all the eigenvalues of* $A(\lambda_0)$ *are distinct, then for an eigenvalue* μ_i *with right and left eigenvectors* $\boldsymbol{x}_i$, $\boldsymbol{y}_i$ *we have*

$$\mu_i^{(2)}(\lambda_0) = \boldsymbol{y}_i' A^{(2)}(\lambda_0)\, \boldsymbol{x}_i + 2 \sum_{\substack{k=1 \\ k \neq i}}^{n} \frac{p_{ik} p_{ki}}{\mu_i - \mu_k} \qquad (2.4.12)$$

where

$$p_{jk} = \boldsymbol{y}_j' A^{(1)}(\lambda_0)\, \boldsymbol{x}_k, \qquad j, k = 1, 2, \ldots, n. \qquad (2.4.13)$$

The assumption that $A(\lambda_0)$ should have all its eigenvalues distinct is relaxed in the next theorem, but at the expense of a further assumption.

THEOREM 2.7 *If, under the hypothesis of Theorem 2.5, we also assume that the matrix* $Y_\alpha' A^{(1)}(\lambda_0)\, X_\alpha$ *is of simple structure, then the eigenvectors of* $\mu(\lambda_0)$ *can be defined in such a way that* (2.4.12) *holds for* $\mu(\lambda_0)$, $\boldsymbol{x}_i$, $\boldsymbol{y}_i$ *being vectors in* $\mathscr{X}_1$, $\mathscr{Y}_1$ *for which* $\boldsymbol{y}_i'\boldsymbol{x}_i = 1$.

We enlarge on this result to point out how X_α, Y_α must be defined in order that Theorem 2.7 will hold. We write

$$P_{11} = Y_\alpha' A^{(1)}(\lambda_0)\, X_\alpha \qquad (2.4.14)$$

and note that, as in Theorem 2.2, X_α, Y_α are not defined to within an arbitrary transformation G applied so that $X_\alpha \to X_\alpha G$, $Y_\alpha \to Y_\alpha (G^{-1})'$. Thus, having found biorthogonal bases which determine X_α, Y_α, we then transform to new bases $X_\alpha G$, $Y_\alpha (G^{-1})'$ where G is determined so that $G^{-1} P_{11}\, G$ is a diagonal matrix, this determination being possible if and only if P_{11} has simple structure (cf. Theorems 1.4 and 1.5). With this choice of G the troublesome terms of (2.4.12), which arise as eigenvalues converge, are eliminated.

Let us illustrate the last theorem with a simple counterexample. Consider the matrix

$$A(\lambda) = \begin{bmatrix} 0 & \lambda \\ \lambda^2 & 0 \end{bmatrix}.$$

The eigenvalues are given by $\mu_{1,2} = \pm \lambda^{3/2}$. They coincide at $\lambda = 0$, and the matrix $A(\lambda)$ has simple structure at $\lambda = 0$. We see that, as Theorem 2.5 implies, the eigenvalues are differentiable at $\lambda = 0$. However,

$$A^{(1)}(\lambda_0) = \begin{bmatrix} 0 & 1 \\ 0 & 0 \end{bmatrix},$$

which does not have simple structure, and the eigenvalues do not have second derivatives at $\lambda = 0$.

2.5 The Constituent Matrices

We consider a set of constituent matrices of a matrix A of simple structure, defined by (2.4.5) and having properties (2.4.6) and (2.4.7). Suppose the eigenvalue μ_1 of A has multiplicity α, so that $\mu_1 = \mu_2 = \cdots = \mu_\alpha$ and we write

$$H_1 = G_1 + G_2 + \cdots + G_\alpha,$$

a matrix of rank α. Let all the constituent matrices be summed in this way according to the multiplicities of the eigenvalues. If there are s distinct eigenvalues, then (2.4.4) and (2.4.10) can be written:

$$(A - \mu I)^{-1} = \sum_{i=1}^{s} \frac{H_i}{\mu_i - \mu} \tag{2.5.1}$$

and

$$p(A) = \sum_{i=1}^{s} p(\mu_i)\, H_i. \tag{2.5.2}$$

The matrices H_i, thus defined, are also idempotent and mutually orthogonal. For, if we take H_1 to be typical, we have

$$\begin{aligned} H_1^2 &= (G_1 + G_2 + \cdots + G_\alpha)^2 \\ &= G_1^2 + G_2^2 + \cdots + G_\alpha^2, && \text{using (2.4.6)}, \\ &= G_1 + G_2 + \cdots + G_\alpha, && \text{using (2.4.7)}. \end{aligned}$$

Hence $H_1^2 = H_1$ and H_1 is idempotent. If H_2 is another idempotent matrix and

$$H_2 = G_{\alpha+1} + \cdots + G_{\alpha+\beta},$$

then

$$H_1 H_2 = (G_1 + \cdots + G_\alpha)(G_{\alpha+1} + \cdots + G_{\alpha+\beta});$$

and using (2.4.6) we have immediately $H_1 H_2 = 0$. More generally we have

$$\left.\begin{aligned} H_i H_j &= 0. \quad i \neq j, \\ H_i^2 &= H_i, \end{aligned}\right\} \quad 1 \leq i, j \leq s. \tag{2.5.3}$$

In some contexts it is useful to be able to write the "new" constituent matrices H_i in terms of the adjoint of $(A - \mu I)$. In what follows, the rth derivative of a function $f(\mu)$ with respect to μ is denoted by $f^{(r)}(\mu)$, and the value of this derivative at $\mu = \mu_i$ will be written $f^{(r)}(\mu_i)$.

Lemma *Let α_r be the multiplicities of the distinct eigenvalues μ_r $(r = 1, 2, \ldots, s)$ of a matrix A of simple structure. Let $c(\mu)$ be the* *characteristic polynomial of A (eqn.* (1.7.2)); *then the first non-vanishing derivative* $c^{(p)}(\mu_i)$ *is* $c^{(\alpha_i)}(\mu_i)$ *and*

$$c^{(\alpha_i)}(\mu_i) = (-1)^{\alpha_i}(\alpha_i)! \prod_{\substack{r=1 \\ r \neq i}}^{s} (u_r - \mu_i)^{\alpha_r}. \tag{2.5.4}$$

The proof of this lemma is elementary and will not be given here. It follows from the property:

$$c(\mu) = \prod_{i=1}^{s} (\mu_i - \mu)^{\alpha_i}. \tag{2.5.5}$$

THEOREM 2.8 *If A is a matrix of simple structure and* $F(\mu)$ *is the adjoint of* $(A - \mu I)$, *then the idempotent matrix* H_i *of the* α-*fold eigenvalue* μ_i *is given by*

$$H_i = -\frac{\alpha F^{(\alpha-1)}(\mu_i)}{c^{(\alpha)}(\mu_i)}. \tag{2.5.6}$$

Proof From (1.3.3) and (1.7.2) we obtain

$$(A - \mu I)\, F(\mu) = c(\mu)\, I$$

and, if $\mu \neq \mu_i$,

$$F(\mu) = (A - \mu I)^{-1}\, c(\mu).$$

By Leibniz's rule for repeated differentiation:

$$F^{(p)}(\mu) = [(A - \mu I)^{-1}]^{(p)}\, c(\mu) + \binom{p}{1} [(A - \mu I)^{-1}]^{(p-1)}\, c^{(1)}(\mu) + \cdots$$

$$+ \binom{p}{p-1} [(A - \mu I)^{-1}]^{(1)}\, c^{(p-1)}(\mu) + (A - \mu I)^{-1}\, c^{(p)}(\mu).$$

From (2.5.1) we obtain

$$F^{(p)}(\mu) = \sum_{i=1}^{s} \Bigg\{ c(\mu) \frac{p!\, H_i}{(\mu_i - \mu)^{p+1}} + \binom{p}{1} c^{(1)}(\mu) \frac{(p-1)!\, H_i}{(\mu_i - \mu)^{p}} + \cdots$$

$$+ \binom{p}{p-1} c^{(p-1)}(\mu) \frac{H_i}{(\mu_i - \mu)^2} + c^{(p)}(\mu) \frac{H_i}{\mu_i - \mu} \Bigg\}. \tag{2.5.7}$$

If the derivatives $c^{(t)}(\mu)$ are formed from (2.5.5), then, for a β-fold eigenvalue μ_i with $\beta > t$, $c^{(t)}(\mu)$ contains just one term with a factor $(\mu_i - \mu)^{\beta-t}$ and all other terms contain higher powers of $(\mu_i - \mu)$. The term in question is explicitly

$$(-1)^t \frac{\beta!}{(\beta - t)!} (\mu_i - \mu)^{\beta-t} \prod_{\substack{r=1 \\ r \neq 1}}^{s} (\mu_r - \mu)^{\alpha_r}.$$

Thus, if we substitute for the derivatives of $c(\mu)$ in (2.5.7) and then put $\mu = \mu_i$ *where* μ_i *has multiplicity* $p + 1$, we obtain

$$F^{(p)}(\mu_i) = \Big\{p! - \binom{p}{1}(p-1)!\frac{(p+1)!}{p!} + \binom{p}{2}(p-2)!\frac{(p+1)!}{(p-1)!} - \cdots + (-1)^{p-1}\binom{p}{p-1}\frac{(p+1)!}{2!} + (-1)^p(p+1)!\Big\} H_i \prod_{\substack{r=1 \\ r\neq i}}^{s} (u_r - \mu_i)^{\alpha_r}.$$

The summation involved here is elementary and reduces to $(-1)^p p!$ so that substitution from the lemma for the product gives

$$F^{(p)}(\mu_i) = (-1)^p p! H_i (-1)^{p+1} \frac{c^{(p+1)}(\mu_i)}{(p+1)!},$$

and putting $\alpha = p + 1$ gives the result required.

Note that the substitution $\mu = \mu_i$ in (2.5.7) is permissible because that equation simply represents the equality of polynomials of degree not exceeding $n - 1$. Since there are more than n values of μ at which the relation is satisfied it must be true for *all* values of μ.

2.6 Conditions for a Regular Pencil to be Simple

We now return to the concept of a regular matrix pencil in order to formulate a necessary and sufficient condition under which a regular pencil is also a simple pencil. We first prove the following result:

THEOREM 2.9 *Latent vectors corresponding to distinct latent roots of a regular matrix pencil are linearly independent.*

Proof Let $\lambda_1, \lambda_2, \ldots, \lambda_s$ be distinct latent roots of the regular pencil $A\lambda + C$, and let $\boldsymbol{q}_1, \boldsymbol{q}_2, \ldots, \boldsymbol{q}_s$ be right latent vectors, one corresponding to each of the listed latent roots. Suppose there exist constants $c_1, c_2, \ldots, c_s$ such that

$$c_1\boldsymbol{q}_1 + c_2\boldsymbol{q}_2 + \cdots + c_s\boldsymbol{q}_s = \boldsymbol{0}. \tag{2.6.1}$$

Premultiply this equation by $A\lambda_1 + C$, noting that $(A\lambda_1 + C)\boldsymbol{q}_1 = \boldsymbol{0}$; then

$$c_2(A\lambda_1 + C)\boldsymbol{q}_2 + \cdots + c_s(A\lambda_1 + C)\boldsymbol{q}_s = \boldsymbol{0}.$$

Now put $C\boldsymbol{q}_i = -\lambda_i A\boldsymbol{q}_i$, $i = 2, 3, \ldots, s$, and remembering that A is non-singular we find that

$$c_2(\lambda_1 - \lambda_2)\,\boldsymbol{q}_2 + \cdots + c_s(\lambda_1 - \lambda_s)\,\boldsymbol{q}_s = \boldsymbol{0}.$$

Now premultiply by $A\lambda_2 + C$ and it is found that

$$c_3(\lambda_1 - \lambda_3)(\lambda_2 - \lambda_3)\,\boldsymbol{q}_3 + \cdots + c_s(\lambda_1 - \lambda_s)(\lambda_2 - \lambda_s)\,\boldsymbol{q}_s = \boldsymbol{0}.$$

After a sequence of $(s - 1)$ such operations we arrive at

$$c_s(\lambda_1 - \lambda_2)(\lambda_2 - \lambda_s)\cdots(\lambda_{s-1} - \lambda_s)\,\boldsymbol{q}_s = \boldsymbol{0},$$

which implies that $c_s = 0$. But the ordering of the vectors in (2.6.1) is arbitrary, so we can also prove that $c_1 = c_2 = \cdots = c_s = 0$. Thus (2.6.1) implies that $c_i = 0$, $i = 1, 2, \ldots, s$, and the theorem is proved.

Corollary *A matrix pencil of order n having n distinct latent roots is a simple matrix pencil.*

The fact that there are n latent roots shows that the coefficient of λ^n in $|A\lambda + C|$ does not vanish, i.e., that $|A|$ does not vanish. Therefore the matrix pencil is regular. The result now follows from the theorem.

Theorem 2.10 *A regular matrix pencil $A\lambda + C$ is simple if and only if for every latent root λ_i the matrix $A\lambda_i + C$ has degeneracy equal to the multiplicity of λ_i.*

Proof We assume first of all that there is one root, λ_1, with multiplicity $\alpha > 1$, and the remaining roots $\lambda_{\alpha+1}, \lambda_{\alpha+2}, \ldots, \lambda_n$ are distinct. Assume also that $A\lambda_1 + C$ has degeneracy α so that there exist α linearly independent right latent vectors associated with λ_1; $\boldsymbol{q}_1, \boldsymbol{q}_2, \ldots, \boldsymbol{q}_\alpha$, say. Consider a set of constants $c_1, c_2, \ldots, c_n$ such that

$$c_1\boldsymbol{q}_1 + \cdots + c_\alpha\boldsymbol{q}_\alpha + c_{\alpha+1}\boldsymbol{q}_{\alpha+1} + \cdots + c_n\boldsymbol{q}_n = \boldsymbol{0}.$$

By the method of proof of Theorem 2.9 it can be proved that $c_{\alpha+1} = c_{\alpha+2} = \cdots = c_n = 0$. But the vectors $\boldsymbol{q}_1, \boldsymbol{q}_2, \ldots, \boldsymbol{q}_\alpha$ are known to be linearly independent so that this implies $c_1 = c_2 = \cdots = c_\alpha = 0$ also. Thus $A\lambda + C$ has n linearly independent right latent vectors and is therefore a simple matrix pencil.

Clearly the same argument can be extended to cover cases where there are several multiple roots.

The converse follows from the lemma to Theorem 2.2. We will also need the following result:

THEOREM 2.11 *If A and C are real symmetric matrices and A is positive definite, then $A\lambda + C$ is a simple matrix pencil.*

Proof Since A is real, symmetric, and positive definite it has a real, symmetric, positive definite square root $A^{1/2}$ as defined by eqn. (1.8.8). We may therefore write

$$A\lambda + C = A^{1/2}(I\lambda + B)A^{1/2}, \tag{2.6.2}$$

where $B = A^{-1/2}CA^{-1/2}$, and is clearly a real symmetric matrix. Taking determinants in (2.6.2) it is also clear that the latent roots of $A\lambda + C$ are just the eigenvalues of B with the signs reversed. Further, to each latent vector $\boldsymbol{q}$ of $A\lambda + C$ there corresponds an eigenvector $\boldsymbol{x} = A^{1/2}\boldsymbol{q}$ of B; and conversely, $\boldsymbol{q} = A^{-1/2}\boldsymbol{x}$. Now because B is real and symmetric it has n linearly independent eigenvectors which may be assumed to form an orthonormal system (this can be deduced from the first of eqns. (2.4.2)). Since the latent vectors are obtained from these by the non-singular transformation represented by $A^{-1/2}$, it follows that the latent vectors are also linearly independent. This completes the theorem.

We remarked in the proof that the latent roots of the pencil coincide with the eigenvalues of the real symmetric matrix $-B$, and the eigenvalues are known to be real. We have therefore:

COROLLARY *Under the assumptions of the theorem all the latent roots and latent vectors of the pencil are real.*

It may be imagined that Theorem 2.11 could perhaps be proved under the weaker condition that A is merely non-singular, rather than being positive definite, for $A^{-1/2}$ would still exist though it would no longer necessarily be real. This is not the case, for it can be shown that the right eigenvectors of *any* real matrix (and those which do not have simple structure in particular) coincide with the latent vectors of some real, symmetric, regular pencil. This follows from a theorem of Taussky and Zassenhaus [1] which implies that for every real matrix B there is a real, non-singular, symmetric matrix A such that

$$B = A^{-1}B'A, \tag{2.6.3}$$

i.e., such that

$$AB = B'A,$$

and

$$(AB)' = B'A = AB.$$

Hence, if we write $AB = C$, C is a real symmetric matrix. Now premultiply the equation

$$(I\lambda - B)\,\boldsymbol{q} = \boldsymbol{0}$$

by A and we find that

$$(A\lambda + C)\,\boldsymbol{q} = \boldsymbol{0}.$$

That every latent vector of $A\lambda + C$ is an eigenvector of B is apparent on premultiplying the final equation by A^{-1}. Thus, the existence of matrices which are not of simple structure (or which are defective) implies the existence of real, regular, symmetric pencils which are not simple.

2.7 Geometric Implications of the Jordan Canonical Form

Let A be a matrix which does not have simple structure. Referring to Theorem 1.5 the following characterization of A can be deduced (Lancaster [5]):

Theorem 2.12 *To an eigenvalue λ of A for which the multiplicity exceeds the index there corresponds at least one right (left) eigenvector $\boldsymbol{x}$ such that $\boldsymbol{y}'\boldsymbol{x} = 0$ for every vector $\boldsymbol{y}$ in the subspace of left (right) eigenvectors of λ.*

That is, there is a right eigenvector $\boldsymbol{x}$ of λ which is orthogonal to the subspace $\mathscr{Y}_1$ of left eigenvectors of λ, and vice versa. We deduced from (2.4.2) that, for a matrix of simple structure, there exist biorthogonal bases in $\mathscr{X}_1, \mathscr{Y}_1$; but the above result shows that this is not possible if A is not simple.

We may illustrate the theorem with two interesting special cases.

(a) If $|A - \lambda I| = 0$ and $(A - \lambda I)$ has rank $n - 1$, (n being the order of A), then the index of the eigenvalue is 1 and the multiplicity of λ is one or more according as $\boldsymbol{y}'\boldsymbol{x} \neq 0$ or $\boldsymbol{y}'\boldsymbol{x} = 0$, where $\boldsymbol{x}$, $\boldsymbol{y}$ are right and left eigenvectors of λ respectively.

(b) If A is a complex symmetric matrix which is not simple then it has at least one eigenvector $\boldsymbol{x}$ such that $\boldsymbol{x}'\boldsymbol{x} = 0$.

The above theorem can easily be extended to the case of the regular matrix pencil $A\lambda + C$. Let $B = A^{-1}C$, then

$$(A\lambda + C)\boldsymbol{q} = \boldsymbol{0} \qquad \text{implies} \qquad (I\lambda - B)\boldsymbol{q} = \boldsymbol{0},$$

and the right latent vectors of $A\lambda + C$ coincide with the right eigenvectors of B. Clearly $A\lambda + C$ is simple if and only if B has simple structure. Let $\boldsymbol{s}$ be any left eigenvector of B with eigenvalue λ; then

$$\boldsymbol{s}'(I\lambda - B) = \boldsymbol{0}',$$

or,

$$\boldsymbol{s}'A^{-1}(A\lambda + C) = \boldsymbol{0}'.$$

Thus $\boldsymbol{r}' = \boldsymbol{c}'A^{-1}$ is a left latent vector of $A\lambda + C$. Now, if the multiplicity of λ exceeds its index, then there exists a right eigenvector $\boldsymbol{q}$ of B such that $\boldsymbol{s}'\boldsymbol{q} = 0$ for all left eigenvectors $\boldsymbol{s}$ of λ. Then writing $\boldsymbol{s}' = \boldsymbol{r}'A$ we see that, if $\boldsymbol{r}'A\boldsymbol{q} = 0$ for all left latent vectors $\boldsymbol{r}$ of λ, then $A\lambda + C$ is not a simple pencil. In this manner we arrive at:

THEOREM 2.13 *The regular pencil $A\lambda + C$ is defective if and only if there exists a latent root λ with a right latent vector $\boldsymbol{q}$ such that $\boldsymbol{r}'A\boldsymbol{q} = 0$ for all left latent vectors, $\boldsymbol{r}$ of λ.*

2.8 THE RAYLEIGH QUOTIENT

As we will see in Chapter 7, the formulation of the problem of small oscillations of a system about a position of equilibrium gives rise to a latent root problem involving a *symmetric* matrix pencil. Lord Rayleigh ([1], § 88) investigated this problem and obtained the expression

$$\lambda_i = -\frac{\boldsymbol{q}_i'C\boldsymbol{q}_i}{\boldsymbol{q}_i'A\boldsymbol{q}_i}, \tag{2.8.1}$$

giving a latent root in terms of its (not normalized) latent vector. This result follows from Theorem 2.11 and our equation (2.2.3). Rayleigh also suggested an iterative process for improving an approximate evaluation of λ. This process was examined and extended by Temple [1] and Crandall [1], and has been further extended and rigorously analyzed by Ostrowski [2]. Rayleigh's and related iterative processes depend on the fact that the quotient

$$R(\boldsymbol{q}) = -\frac{\boldsymbol{q}'C\boldsymbol{q}}{\boldsymbol{q}'A\boldsymbol{q}}, \tag{2.8.2}$$

considered as a function of the components of $\boldsymbol{q}$, has a stationary value when $\boldsymbol{q}$ is a latent vector of $A\lambda + C$. Full accounts of this and related properties can be found in Gould [1] and Gantmacher [1], Chapter 10. For the unsymmetric case the work of Ostrowski suggests that we consider the quotient

$$R(\boldsymbol{q}, \boldsymbol{r}) = -\frac{\boldsymbol{r}'C\boldsymbol{q}}{\boldsymbol{r}'A\boldsymbol{q}}. \tag{2.8.3}$$

This obviously includes (2.8.2) as a special case. The following theorem concerning this quotient can now be proved:

THEOREM 2.14 *If $A\lambda + C$ is a simple pencil of matrices, and the latent vectors are defined in such a way that* (2.2.1) *is true, then $R(\boldsymbol{q}, \boldsymbol{r})$ has a stationary value at $\boldsymbol{q} = \boldsymbol{q}_i$, $\boldsymbol{r} = \boldsymbol{r}_i$ (a pair of latent vectors associated with the latent root λ_i) for $i = 1, 2, \ldots, n$; and $R(\boldsymbol{q}_i, \boldsymbol{r}_i) = \lambda_i$.*

The last part of the theorem follows immediately from eqns. (2.2.3). We will not prove the stationary property here as this theorem will arise later as a special case of a more general result which we *will* prove. Notice that the *existence* of R for a simple pencil is assured by the result of Theorem 2.13.

For the pencils of matrices discussed in Theorem 2.11, the Rayleigh quotient has another important property. This is an extension of the result (1.7.3). We have seen in the corollary to Theorem 2.11 that the latent roots $\lambda_1, \lambda_2, \ldots, \lambda_n$ are real. Suppose that they are arranged in non-decreasing order so that $\lambda_1 \leqq \lambda_2 \leqq \ldots \leqq \lambda_n$. We have seen that these numbers are also the eigenvalues of the real symmetric matrix $B = -A^{-1/2}CA^{-1/2}$, and the result (1.7.3) now implies that

$$\lambda_1 \leqq -\bar{\boldsymbol{x}}'A^{-1/2}CA^{-1/2}\boldsymbol{x} \leqq \lambda_n,$$

for any $\boldsymbol{x}$ in $\mathscr{C}_n$ such that $\bar{\boldsymbol{x}}'\boldsymbol{x} = 1$. Thus, writing $\boldsymbol{y} = A^{-1/2}\boldsymbol{x}$ we have

$$\lambda_1 \leqq -\bar{\boldsymbol{y}}'C\boldsymbol{y} \leqq \lambda_n$$

for any $\boldsymbol{y}$ in $\mathscr{C}_n$ such that $\bar{\boldsymbol{y}}'A\boldsymbol{y} = 1$. These inequalities imply that

$$\lambda_1 \leqq -\frac{\bar{\boldsymbol{y}}'C\boldsymbol{y}}{\bar{\boldsymbol{y}}'A\boldsymbol{y}} \leqq \lambda_n \tag{2.8.4}$$

for any $\boldsymbol{y} \neq 0$ in $\mathscr{C}_n$, since the quotient is homogeneous in $\boldsymbol{y}$. If we restrict $\boldsymbol{y}$ to belong to $\mathscr{R}_n$, then we arrive at the Rayleigh quotient (2.8.2).

2.9 Simple Matrix Pencils with Latent Vectors in Common

Theorem 2.1 can easily be modified so that eqns. (2.2.1) appear in the form

$$R'AQ = D \qquad \text{and} \qquad R'CQ = -D\Lambda \tag{2.9.1}$$

where D is an arbitrary non-singular diagonal matrix and $A\lambda + C$ is a simple pencil of matrices. Later on we will need to consider matrix pencils related to this by having the same latent vectors but different latent roots.

For future convenience we formulate the problem in the following way: If Λ is arbitrarily divided into two diagonal matrices so that

$$\Lambda = \Lambda_0 + \Delta, \tag{2.9.2}$$

say, what matrix pencil has the matrices of latent vectors Q and R and the diagonal matrix of latent roots Λ_0?

Suppose the pencil with the required properties is $A_1\lambda + C_1$. Then as in (2.9.1) we have

$$R'A_1Q = D_1 \qquad \text{and} \qquad R'C_1Q = -D_1\Lambda_0 \tag{2.9.3}$$

for some diagonal matrix D_1.

By choice of D and D_1 we now define two matrix pencils with the required properties.

Case 1 If Λ_0 is assumed to be non-singular we may choose $D = I$ and $D_1 = (\Lambda_0 + \Delta)\,\Lambda_0^{-1}$; then using (2.9.2), eqns. (2.9.1) and (2.9.3) become

$$\begin{aligned} R'AQ &= I, & R'CQ &= -(\Lambda_0 + \Delta), \\ R'A_1Q &= I + \Delta\Lambda_0{}^{-1}, & R'C_1Q &= -(\Lambda_0 + \Delta). \end{aligned} \tag{2.9.4}$$

Thus we obtain the solution

$$A_1 = A + (R')^{-1}\Delta^{-1}Q^{-1}, \quad C_1 = C. \tag{2.9.5}$$

Case 2 Choose $D = D_1 = I$, and it is easily found that

$$A_1 = A \qquad \text{and} \qquad C_1 = C - (R')^{-1}\Delta Q^{-1} \tag{2.9.6}$$

provide an alternative solution.

CHAPTER 3

LAMBDA-MATRICES, I

3.1 Introduction

We now consider an $m \times n$ matrix whose elements are polynomials in a scalar, λ; the coefficients of the polynomials may be complex numbers. Such a matrix may obviously be considered as a polynomial in λ whose coefficients are constant $m \times n$ matrices. If l is the highest power of λ appearing among the elements, we refer to such a matrix as a λ-matrix of degree l, and write

$$D_l(\lambda) = A_0\lambda^l + A_1\lambda^{l-1} + \cdots + A_{l-1}\lambda + A_l, \qquad (3.1.1)$$

where $A_0, A_1, \ldots, A_l$ are $m \times n$ matrices with elements from the field of complex numbers.

We recall (§ 1.6) that the latent roots λ_i and the right and left latent vectors $\boldsymbol{q}_i, \boldsymbol{r}_i$ of $D_l(\lambda)$ are defined so that

$$D_l(\lambda_i)\,\boldsymbol{q}_i = \boldsymbol{0} \qquad \text{and} \qquad \boldsymbol{r}_i'D_l(\lambda_i) = \boldsymbol{0}'. \qquad (3.1.2)$$

A λ-matrix is said to be *regular* if and only if it is a square matrix and the matrix of coefficients A_0 in (3.1.1) is non-singular.

We define the *latent equation* of a square λ-matrix $D_l(\lambda)$ to be

$$\Delta(\lambda) = |D_l(\lambda)| = 0. \qquad (3.1.3)$$

This is a polynomial equation in λ whose degree does not exceed ln, and the *multiplicity* of a latent root λ_i is defined to be the number of times the factor $(\lambda - \lambda_i)$ appears in the factorization of $\Delta(\lambda)$ into linear factors. The coefficient of λ^{ln} in (3.1.3) is $|A_0|$ so that a regular λ-matrix must have ln latent roots if they are counted according to their multiplicities.

We next define a *simple* λ-matrix to be one which:

(a) Is a regular λ-matrix.

(b) Has degeneracy α (or rank $n - \alpha$) when evaluated at $\lambda = \lambda_i$, where λ_i is a latent root of multiplicity α.

Since the degeneracy of an $n \times n$ matrix cannot exceed n, we immediately deduce that the multiplicity of any latent root of a simple λ-matrix cannot exceed n. We also note that the definition implies the existence of α *linearly independent* vectors $\boldsymbol{q}_i$ associated with λ_i, and similarly for the $\boldsymbol{r}_i$ (cf. eqns. (3.1.2)). The corresponding property for matrix pencils is described in the lemma to Theorem 2.2. A square λ-matrix is *defective* if there exists a latent root whose multiplicity exceeds the dimension of the subspace of its right latent vectors.

The *rank* of a λ-matrix is the order of its largest minor which does not vanish *identically*. In this sense, the rank of a regular λ-matrix is equal to its order; for, if not, then $|D_l(\lambda)| \equiv 0$, which implies that the coefficient of each power of λ in (3.1.3) must vanish. In particular $|A_0| = 0$ (being the coefficient of λ^{ln}), which is not the case for a regular λ-matrix.

The purpose of this chapter is, first, to introduce a canonical form for regular λ-matrices followed by the notion of elementary divisors. We then obtain an alternative definition of simple λ-matrices from the properties of the elementary divisors. The last half of the chapter is concerned with matrix polynomials with a matrix argument, i.e., of the form $D_l(A)$. We consider the possibility of factorizing $D_l(A)$ and of finding solutions of $D_l(A) = 0$.

3.2 A Canonical Form for Regular λ-Matrices

We introduce the following right and left elementary operations on a square λ-matrix.† The words in parentheses will refer to the left elementary operations:

(1) Interchange of two columns (rows).

(2) Multiplication of any column (row) by a number $c \neq 0$.

(3) Addition to any column (row) of any other column (row) multiplied by an arbitrary scalar polynomial $b(\lambda)$.

It is easily verified that postmultiplication of a 5×5 λ-matrix by each of the following matrices is equivalent to a right elementary operation:

† These operations and the idea of equivalence are easily extended to rectangular λ-matrices.

$$S_1 = \begin{bmatrix} 1 & 0 & 0 & 0 & 0 \\ 0 & 0 & 0 & 1 & 0 \\ 0 & 0 & 1 & 0 & 0 \\ 0 & 1 & 0 & 0 & 0 \\ 0 & 0 & 0 & 0 & 1 \end{bmatrix}, \quad S_2 = \begin{bmatrix} 1 & 0 & 0 & 0 & 0 \\ 0 & c & 0 & 0 & 0 \\ 0 & 0 & 1 & 0 & 0 \\ 0 & 0 & 0 & 1 & 0 \\ 0 & 0 & 0 & 0 & 1 \end{bmatrix}, \quad S_3 = \begin{bmatrix} 1 & 0 & 0 & 0 & 0 \\ 0 & 1 & 0 & 0 & 0 \\ 0 & 0 & 1 & 0 & 0 \\ 0 & b(\lambda) & 0 & 1 & 0 \\ 0 & 0 & 0 & 0 & 1 \end{bmatrix}.$$

With S_1 we interchange the second and fourth columns; with S_2 the second column is multiplied by c; and with S_3 the fourth column multiplied by $b(\lambda)$ is added to the second column. S_1, S_2, S_3 are there fore known as *right elementary matrices*. For an $n \times n$ λ-matrix the corresponding right elementary matrices giving operations on the ith and jth columns are easily formulated.

The analogous *left elementary matrices* are found to be given by

$$T_1 = S_1, \qquad T_2 = S_2, \qquad \text{and} \qquad T_3 = S_3',$$

but we must *pre*multiply the λ-matrix by T_1, T_2, T_3 to get equivalence with the left elementary operations.

We note that all the elementary matrices have non-zero determinants and that their determinants are independent of λ. These properties will obviously carry over to all products of the elementary matrices among themselves. We observe also that the inverse of a right (left) elementary matrix is itself a right (left) elementary matrix.

Two λ-matrices are said to be equivalent if one can be obtained from the other by means of left and right elementary operations. It can be proved that any square λ-matrix with a constant non-zero determinant can be expressed as a product of elementary matrices. Thus, square λ-matrices $A(\lambda)$ and $B(\lambda)$ of the same order are equivalent if and only if there exist λ-matrices $E(\lambda)$ and $F(\lambda)$ with the property that $|E(\lambda)|$ and $|F(\lambda)|$ are non-zero constants and

$$B(\lambda) = E(\lambda)\,A(\lambda)\,F(\lambda). \tag{3.2.1}$$

We may now state an important theorem concerning canonical forms. Proofs are given by Ferrar [1], Gantmacher [1], Perlis [1], and Turnbull and Aitken [1]. Having established in § 3.1 that a regular λ-matrix of order n has rank n, we may formulate the theorem as follows:

THEOREM 3.1 *A regular λ-matrix $D_l(\lambda)$ can be reduced by a finite sequence of elementary operations to a diagonal matrix,*

$$\Gamma(\lambda) = \begin{bmatrix} a_1(\lambda) & 0 & \dots & 0 \\ 0 & a_2(\lambda) & \dots & 0 \\ \vdots & \vdots & & \vdots \\ 0 & 0 & \dots & a_n(\lambda) \end{bmatrix} = E(\lambda)\, D_l(\lambda)\, F(\lambda), \tag{3.2.2}$$

where: (a) $a_1(\lambda), \dots, a_n(\lambda)$ *are monic polynomials in* λ.

(b) *The polynomials* $a_1(\lambda), \dots, a_n(\lambda)$ *are unique for a given* λ*-matrix.*

(c) *Each of the polynomials* $a_2(\lambda), \dots, a_n(\lambda)$ *is divisible by the previous one.*

The polynomials $a_1(\lambda), \dots, a_n(\lambda)$ are known as the invariant polynomials of $D_l(\lambda)$. It follows as an immediate corollary of this theorem that regular λ-matrices are equivalent if and only if they have the same invariant polynomials.

3.3 ELEMENTARY DIVISORS

We now proceed to write the invariant polynomials as products of their (possibly multiple) linear factors. This is always possible if we consider the factors with coefficients in the field of complex numbers. Since $|E(\lambda)|$ and $|F(\lambda)|$ are non-zero constants we obtain from (3.2.2):

$$|\Gamma(\lambda)| = \prod_{i=1}^{n} a_i(\lambda) = (\text{const})\, |D_l(\lambda)|.$$

Hence the zeros of the invariant polynomials coincide with the latent roots of $D_l(\lambda)$ and conversely. Hence, using property (a) of the invariant polynomials:

$$\prod_{i=1}^{n} a_i(\lambda) = \prod_{i=1}^{ln} (\lambda - \lambda_i),$$

where the λ_i are the latent roots of $D_l(\lambda)$. If there are *s distinct* latent roots $\lambda_1, \lambda_2, \dots, \lambda_s$, we may now write

$$\left.\begin{aligned} a_1(\lambda) &= (\lambda - \lambda_1)^{\alpha_{1,1}} (\lambda - \lambda_2)^{\alpha_{1,2}} \cdots (\lambda - \lambda_s)^{\alpha_{1,s}} \\ a_2(\lambda) &= (\lambda - \lambda_1)^{\alpha_{2,1}} (\lambda - \lambda_2)^{\alpha_{2,2}} \cdots (\lambda - \lambda_s)^{\alpha_{2,s}} \\ &\vdots \\ a_n(\lambda) &= (\lambda - \lambda_1)^{\alpha_{n,1}} (\lambda - \lambda_2)^{\alpha_{n,2}} \cdots (\lambda - \lambda_s)^{\alpha_{n,s}} \end{aligned}\right\} \tag{3.3.1}$$

where the $\alpha_{i,j}$ $(1 \leqq i \leqq n,\ 1 \leqq j \leqq s)$ are non-negative integers such that

$$\sum\sum \alpha_{ij} = ln. \tag{3.3.2}$$

Using property (c) of the invariant polynomials we also have:

$$0 \leqq \alpha_{1,j} \leqq \alpha_{2,j} \leqq \cdots \leqq \alpha_{n,j}. \tag{3.3.3}$$

All the factors $(\lambda - \lambda_j)^{\alpha_{ij}}$ appearing in (3.3.1) for which the exponent $\alpha_{i,j} \neq 0$ are called the *elementary divisors* of $D_l(\lambda)$. An elementary divisor with exponent 1 is said to be linear. We also refer to the elementary divisors of λ_j, with the obvious meaning.

Clearly, if there are *ln distinct* latent roots, then

$$a_1(\lambda) = a_2(\lambda) = \cdots = a_{n-1}(\lambda) = 1, \qquad a_n(\lambda) = \prod_{i=1}^{ln} (\lambda - \lambda_i),$$

and all the elementary divisors are linear. We notice also that, if a root of multiplicity x has only linear elementary divisors, then we must have $x \leqq n$.

THEOREM 3.2 *Let $D_l(\lambda)$ be a regular λ-matrix with a latent root λ_j of multiplicity x where $1 \leqq x \leqq n$, then λ_j has only linear elementary divisors if and only if the (constant) matrix $D_l(\lambda_j)$ has degeneracy x.*

Proof From (3.2.2) we see that $D_l(\lambda_j)$ and $\Gamma(\lambda_j)$ have the same rank, because $E(\lambda_j)$ and $F(\lambda_j)$ are non-singular. Thus, if $D_l(\lambda_j)$ has degeneracy x, then so has $\Gamma(\lambda_j)$, and this can occur only if λ_j is a zero of x of the invariant polynomials. By (3.3.3) λ_j must be a zero of $a_{n-x+1}, \ldots, a_n$, and since x is also the multiplicity of λ_j we have in (3.3.1):

$$\alpha_{1,j} = \alpha_{2,j} = \cdots = \alpha_{n-x,j} = 0,$$

and

$$\alpha_{n-x+1,j} = \cdots = \alpha_{n-1,j} = \alpha_{n,j} = 1, \tag{3.3.4}$$

so that all the elementary divisors of λ_j are linear.

Conversely, given eqns. (3.3.4) the last x of the invariant polynomials have zeros at $\lambda = \lambda_j$ so that $\Gamma(\lambda_j)$ has degeneracy x, and hence $D_l(\lambda_j)$ has degeneracy x. This completes the proof.

Combining this theorem with the definition of a simple λ-matrix we immediately obtain:

COROLLARY 1 *A regular λ-matrix is simple if and only if all its elementary divisors are linear.*

If A is a constant matrix, then the latent roots of $I\lambda - A$ are the eigenvalues of A. In the sequel, the "elementary divisors of an eigenvalue of A" will imply the appropriate elementary divisors of $I\lambda - A$. In this connection, we may also infer from theorem 3.2:

COROLLARY 2 *The matrix A has simple structure if and only if all the elementary divisors of $I\lambda - A$ are linear.*

The following characterizations may also be obtained by combining earlier results:

COROLLARY 3 *An eigenvalue of A has only linear elementary divisors if and only if:*

(i) *There exist biorthogonal bases in the subspaces of its right and left eigenvectors. Or, equivalently:*

(ii) *The index of the eigenvalue is equal to its multiplicity.*

3.4 DIVISION OF SQUARE λ-MATRICES

Consider the square λ-matrices $D_l(\lambda)$ and $D_m(\lambda)$, each or order n and with $m \leq l$. Let $D_m(\lambda)$ be regular. We say that $Q(\lambda)$ is a right quotient of $D_l(\lambda)$ and $R(\lambda)$ is the right remainder of $D_l(\lambda)$ on division by $D_m(\lambda)$ if

$$D_l(\lambda) = Q(\lambda)\, D_m(\lambda) + R(\lambda), \tag{3.4.1}$$

provided the degree of $R(\lambda)$ is less than m.

Similarly $\hat{Q}(\lambda)$ and $\hat{R}(\lambda)$ are defined as the left quotient and left remainder of $D_l(\lambda)$ on division by $D_m(\lambda)$ if

$$D_l(\lambda) = D_m(\lambda)\, \hat{Q}(\lambda) + \hat{R}(\lambda), \tag{3.4.2}$$

provided the degree of $\hat{R}(\lambda)$ is less than m.

We will assume that, as defined here, the processes of left and right division of matrix polynomials of the same order is unique. A proof can be found in Chapter 4, § 2, of Gantmacher [1]. It is important to note, however, that the uniqueness can be proved only if the divisor is a regular λ-matrix.†

The definition of right quotient and remainder can easily be extended to dividends $D_l(\lambda)$, which are $p \times n$ where the divisor $D_m(\lambda)$ is a regular $n \times n$ matrix. The uniqueness property is preserved and $Q(\lambda)$, $R(\lambda)$ are also $p \times n$ matrices.

If $p < n$, we need only add $n - p$ rows of zeros to $D_l(\lambda)$, use the above results, and then verify that Q and R have their last $n - p$ rows null. If $p > n$ we first add null rows to $D_l(\lambda)$ until the

† Quotients obtained when the divisor is not regular are investigated by Flood [1].

total number of rows is a multiple of n and then break up the resulting matrix into $n \times n$ partitions to each of which the above result may be applied. Again, Q and R are $p \times n$ matrices.

In a similar manner the definition of left quotient and remainder may be extended to $n \times p$ matrices, resulting in unique λ-matrices $\hat{Q}$ and $\hat{R}$ which are also $n \times p$.

In our definition of a λ-matrix, or matrix polynomial, it is immaterial whether we write $D_l(\lambda)$ as in equation (3.1.1) or whether we write

$$D_l(\lambda) = \lambda^l A_0 + \lambda^{l-1} A_1 + \cdots + A_l.$$

However, if we consider a matrix polynomial of order n whose argument is a square matrix A of the same order, then the two forms are in general distinct and we will write

$$D_l(A) = A_0 A^l + A_1 A^{l-1} + \cdots + A_l, \tag{3.4.3}$$

and

$$\hat{D}_l(A) = A^l A_0 + A^{l-1} A_1 + \cdots + A_l. \tag{3.4.4}$$

THEOREM 3.3 *When the λ-matrix $D_l(\lambda)$ is divided on the right by $I\lambda - A$ the remainder is $D_l(A)$, and when $D_l(\lambda)$ is divided on the left by $I\lambda - A$ the remainder is $\hat{D}_l(A)$.*

Proof We have

$$\begin{aligned}
D_l(\lambda) &= A_0\lambda^l + A_1\lambda^{l-1} + \cdots + A_l \\
&= A_0\lambda^{l-1}(I\lambda - A) + (A_0A + A_1)\,\lambda^{l-1} + A_2\lambda^{l-2} + \cdots + A_l \\
&= [A_0\lambda^{l-1} + (A_0A + A_1)\,\lambda^{l-2}]\,(I\lambda - A) \\
&\qquad + (A_0A^2 + A_1A + A_2)\,\lambda^{l-2} + A_3\lambda^{l-3} + \cdots + A_l \\
&= [A_0\lambda^{l-1} + (A_0A + A_1)\,\lambda^{l-2} + \cdots \\
&\qquad + (A_0A^{l-1} + A_1A^{l-2} + \cdots + A_{l-1})]\,(I\lambda - A) \\
&\qquad\qquad + A_0A^l + A_1A^{l-1} + \cdots + A_{l-1}A + A_l.
\end{aligned}$$

Comparing this with (3.4.1) and observing that $(I\lambda - A)$ is a λ-matrix of degree one, we see that $D_l(\lambda)$ is divisible on the right by $I\lambda - A$ with remainder

$$R(\lambda) = A_0A^l + A_1A^{l-1} + \cdots + A_l = D_l(A).$$

Similarly, we find that $D_l(\lambda)$ is divisible on the left by $I\lambda - A$ with remainder

$$\hat{R}(\lambda) = \hat{D}_l(A). \qquad \text{Q.E.D.}$$

Corollary *A λ-matrix is divisible on the right (left) by* $I\lambda - A$ *if and only if* $D_l(A) = 0$ $(\hat{D}_l(A) = 0)$.

We will call the matrix A a right solvent of $D_l(\lambda)$ if we have $D_l(A) = 0$, and A is a left solvent of $D_l(\lambda)$ if $\hat{D}_l(A) = 0$. We will now construct examples of such matrices.

Suppose that the λ-matrix $D_l(\lambda)$ possesses n right latent vectors which form a linearly independent set. Call these $\mathbf{q}_1, \mathbf{q}_2, \ldots, \mathbf{q}_n$, write

$$Q_1 = [\mathbf{q}_1, \mathbf{q}_2, \cdots, \mathbf{q}_n],$$

and let Λ_1 be the diagonal matrix of the associated latent roots (not necessarily distinct). With these definitions the columns of the matrix $A_p Q_1 \Lambda_1^p$ $(0 \leqq p \leqq l)$ are $A_p \mathbf{q}_1 \lambda_1^p, \ldots, A_p \mathbf{q}_n \lambda_n^p$. Thus, using (3.1.1) and (3.1.2)

$$A_0 Q_1 \Lambda_1^l + A_1 Q_1 \Lambda_1^{l-1} + \ldots + A_{l-1} Q_1 \Lambda_1 + A_l Q_1 = 0. \qquad (3.4.5)$$

Postmultiply this equation by Q_1^{-1}, note that $Q_1 \Lambda_1^l Q_1^{-1} = (Q_1 \Lambda_1 Q_1^{-1})^l$, etc., and we have

$$D_l(Q_1 \Lambda_1 Q_1^{-1}) = 0. \qquad (3.4.6)$$

Thus, $(Q_1 \Lambda_1 Q_1^{-1})$ is a right solvent of $D_l(\lambda)$.

If the columns of $R_1 = [\mathbf{r}_1, \mathbf{r}_2, \ldots, \mathbf{r}_n]$ are linearly independent left latent vectors of $D_l(\lambda)$ with the diagonal matrix Λ_1 of associated latent roots, then we find similarly that $(R_1 \Lambda_1 R_1^{-1})'$ is a left solvent of $D_l(\lambda)$.

If we write

$$A = Q_1 \Lambda_1 Q_1^{-1}$$

then

$$A Q_1 = Q_1 \Lambda_1,$$

and comparison with (1.8.4) shows that the right eigenvectors of the right solvent A are right latent vectors of $D_l(\lambda)$, and the eigenvalues of A are all latent roots of $D_l(\lambda)$, as is otherwise obvious. A similar result holds for the left eigenvectors of the left solvent $B = (R_1 \lambda_1 R_1^{-1})'$, and for the eigenvalues of B. We will investigate the nature of solvent matrices more closely in § 3.7.

3.5 The Cayley–Hamilton Theorem

The following important result of matrix theory provides an interesting application of Theorem 3.3.

THEOREM 3.4 *Every matrix satisfies its own characteristic equation.* In the notation of (1.7.2) the theorem states that

$$c(A) = 0.$$

Proof Let $F(\lambda)$ be the adjoint of $(A - \lambda I)$, then by definition (§ 1.3) $F(\lambda)$ is a λ-matrix whose degree does not exceed $n - 1$, and (1.3.3) and (1.3.4) give

$$(A - \lambda I)\, F(\lambda) = c(\lambda)\, I,$$

and

$$F(\lambda)\,(A - \lambda I) = c(\lambda)\, I.$$

Comparing these relations with (3.4.1) and (3.4.2) we observe that the λ-matrix $c(\lambda)\, I$ is divisible on the left and on the right by $(A - \lambda I)$ without remainder. The theorem now follows from the corollary to the previous theorem.

3.6 DECOMPOSITION OF λ-MATRICES

In this section we use the ideas developed in § 3.4 to show how certain regular λ-matrices can be decomposed into products of factors which are linear in λ. The complexity of these decompositions increases with the degree l of the λ-matrix, so we restrict our attention to the cases $l = 2$, 3, and 4. No new principles would be involved in formulating similar decompositions of regular λ-matrices with $l > 4$.

THEOREM 3.5 *Let $D_2(\lambda)$ be a regular λ-matrix. Assume that the $2n$ latent roots of $D_2(\lambda)$ can be split into two disjoint sets*

$$\lambda_1, \lambda_2, \ldots, \lambda_n \qquad \textit{and} \qquad \lambda_{n+1}, \lambda_{n+2}, \ldots, \lambda_{2n},$$

and that there are n linearly independent right latent vectors $\boldsymbol{q}_1, \boldsymbol{q}_2, \ldots, \boldsymbol{q}_n$ associated with $\lambda_1, \lambda_2, \ldots, \lambda_n$ and n linearly independent left latent vectors $\boldsymbol{r}_{n+1}, \ldots, \boldsymbol{r}_{2n}$ associated with the second set.

If we write

$$Q_1 = [\boldsymbol{q}_1, \boldsymbol{q}_2, \ldots, \boldsymbol{q}_n], \qquad R_2 = [\boldsymbol{r}_{n+1}, \boldsymbol{r}_{n+2}, \ldots, \boldsymbol{r}_{2n}]$$

and Λ_1, Λ_2 for the associated diagonal matrices of latent roots, then

$$D_2(\lambda) = (I\lambda - (R_2 \Lambda_2 R_2^{-1})')\, A_0 (I\lambda - (Q_1 \Lambda_1 Q_1^{-1})). \quad (3.6.1)$$

Proof We have already seen (in (3.4.6)) that the matrix

$$A = Q_1 \Lambda_1 Q_1^{-1} \quad (3.6.2)$$

is a right solvent of $D_2(\lambda)$. Hence there exists a square matrix B of order n such that

$$D_2(\lambda) = (I\lambda - B)\,A_0(I\lambda - A), \tag{3.6.3}$$

and the eigenvalues of A are $\lambda_1, \lambda_2, \dots, \lambda_n$. Now consider

$$\boldsymbol{r}_i' D_2(\lambda_i) = \boldsymbol{r}_i'(I\lambda_i - B)A_0(I\lambda_i - A) = \boldsymbol{0} \quad \text{for} \quad i = n+1, \cdots, 2n.$$

Since our two sets of latent roots are assumed to be disjoint, the matrix $A_0(I\lambda_i - A)$ is non-singular for each of these values of i, and the previous equation implies that the vectors $\boldsymbol{r}_i$ are the left eigenvectors of B and the second set of latent roots contains all the eigenvalues of B. Hence

$$R_2'B = \Lambda_2 R_2',$$

and R_2 is non-singular by hypothesis, so that

$$B = (R_2 \Lambda_2 R_2^{-1})'. \tag{3.6.4}$$

Substituting from (3.6.4) and (3.6.2) into (3.6.3) proves the theorem. We shall see later (Theorem 4.9) that the conditions made in the statement of this theorem imply that $D_2(\lambda)$ is a simple λ-matrix. Though the hypothesis of the theorem may appear to be very restrictive, we will consider a class of vibration problems (§ 7.6) in which they are always satisfied.

THEOREM 3.6 *Let $D_3(\lambda)$ be a regular λ-matrix and assume that the $3n$ latent roots of $D_3(\lambda)$ can be divided into three mutually disjoint sets*

$$\lambda_1, \lambda_2, \dots, \lambda_n; \qquad \lambda_{n+1}, \lambda_{n+2}, \dots, \lambda_{2n}; \qquad \lambda_{2n+1}, \dots, \lambda_{3n}.$$

If we write

$$Q_{p+1} = [\boldsymbol{q}_{pn+1}, \dots, \boldsymbol{q}_{pn+n}] \qquad R_{p+1} = [\boldsymbol{r}_{pn+1}, \dots, \boldsymbol{r}_{pn+n}] \tag{3.6.5}$$

for $p = 0, 1, 2$, assume that Q_1, Q_3, R_2, R_3 are non-singular, and let $\Lambda_1, \Lambda_2, \Lambda_3$ be the diagonal matrices of the above sets of latent roots, then the matrices of latent vectors can be defined in such a way that

$$D_3(\lambda) = (I\lambda - Y_2)\,(Y_3 - Y_2)^{-1}\,R_3'^{-1}\,(I\lambda - \Lambda_3)\,Q_3^{-1}(X_3 - X_1)^{-1} \times (I\lambda - X_1), \tag{3.6.6}$$

where

$$X_p = Q_p \Lambda_p Q_p^{-1} \qquad \text{and} \qquad Y_p = (R_p \Lambda_p R_p^{-1})', \tag{3.6.7}$$

and it is assumed that $(X_3 - X_1)$ and $(Y_3 - Y_2)$ are non-singular.

Proof We have already observed (in § 3.4) that the matrices X_p and Y_p are right and left solvents of $D_3(\lambda)$ respectively. Using the argument of the previous theorem we may evidently write:

$$D_3(\lambda) = (I\lambda - Y_2)(A_0\lambda + B)(I\lambda - X_1) \qquad (3.6.8)$$

for some square matrix B of order n. Then

$$D_3(\lambda_i)\, \boldsymbol{q}_i = (I\lambda_i - Y_2)(A_0\lambda_i + B)(I\lambda_i - X_1)\, \boldsymbol{q}_i = \boldsymbol{0}$$

for $i = 2n+1, \ldots, 3n$. For these values of i the matrices $(I\lambda_i - Y_2)$ and $(I\lambda_i - X_1)$ are non-singular, so that the right latent vectors of $(A_0\lambda + B)$ are given by

$$\boldsymbol{s}_i = (I\lambda_i - X_1)\, \boldsymbol{q}_i.$$

Thus,

$$S = [\boldsymbol{s}_{2n+1}, \ldots, \boldsymbol{s}_{3n}] = Q_3\Lambda_3 - X_1 Q_3,$$

and using the definitions (3.6.7),

$$S = (X_3 - X_1)\, Q_3. \qquad (3.6.9)$$

Similarly, if T is the matrix of left latent vectors of $(A_0\lambda + B)$, we obtain

$$T' = R_3'(Y_3 - Y_2). \qquad (3.6.10)$$

By hypothesis $(X_3 - X_1)$ and $(Y_3 - Y_2)$ are non-singular so that S and T are also non-singular. It therefore follows that $A_0\lambda + B$ is a simple matrix pencil and using Theorem 2.1 we see that S and T can be defined (by manipulation of Q_3 and R_3) so that

$$T'(A_0\lambda + B)\, S = I\lambda - \lambda_3.$$

Hence

$$\begin{aligned} A_0\lambda + B &= T'^{-1}(I\lambda - \Lambda_3)\, S^{-1} \\ &= (Y_3 - Y_2)^{-1}\, R_3'^{-1}(I\lambda - \Lambda_3)\, Q_3^{-1}(X_3 - X_1)^{-1}, \end{aligned}$$

using (3.6.9) and (3.6.10). Substitution in (3.6.8) now gives the required result.

It should be noted that the factorization, or decomposition, theorems are not unique results. The forms presented are intended to take advantage of whatever symmetries the problem may possess.

In the case $l = 4$ a similar theorem can be proved without too much difficulty. The same sort of assumptions are made as in Theorem 3.6 with definitions analogous to (3.6.5) and (3.6.7). It can then be proved that

$$\begin{aligned} D_4(\lambda) = (I\lambda - Y_2)(Y_4 - Y_2)^{-1}\,(I\lambda - Y_4)(Y_4 - Y_2)\, A_0(X_3 - X_1) \\ \times\, (I\lambda - X_3)(X_3 - X_1)^{-1}\,(I\lambda - X_1). \end{aligned}$$

3.7 Matrix Polynomials with a Matrix Argument

Theorem 3.7 (i) *If A is a right (left) solvent of $D_l(\lambda)$, then every eigenvalue of A is a latent root of $D_l(\lambda)$ and every right (left) eigenvector of A is a right (left) latent vector of $D_l(\lambda)$.*

(ii) *Let A be an $n \times n$ matrix of simple structure whose eigenvalues $\lambda_1, \lambda_2, \ldots, \lambda_n$ coincide with latent roots of the $n \times n$ λ-matrix $D_l(\lambda)$ while a linearly independent set of n associated right (left) eigenvectors of A coincide with latent vectors of $D_l(\lambda)$; then A is a right (left) solvent of $D_l(\lambda)$.*

Proof If $D_l(A) = 0$, then by the corollary to Theorem 3.3 there exists a λ-matrix $Q(\lambda)$ such that

$$D_l(\lambda) = Q(\lambda)(I\lambda - A). \tag{3.7.1}$$

Let

$$\Delta(\lambda) = |D_l(\lambda)|, \tag{3.7.2}$$

then taking determinants in (3.7.1):

$$\Delta(\lambda) = (-1)^n |Q(\lambda)| \, c(\lambda), \tag{3.7.3}$$

where $c(\lambda)$ is the characteristic polynomial of A. Now $c(\lambda) = 0$ if and only if λ is an eigenvalue of A, and $c(\lambda) = 0$ implies that $\Delta(\lambda) = 0$, i.e., every eigenvalue of A is also a latent root of $D_l(\lambda)$.

If $\boldsymbol{x}_i$ is a right eigenvector of A associated with the eigenvalue λ_i, then by definition

$$(I\lambda_i - A)\,\boldsymbol{x}_i = \boldsymbol{0},$$

and from (3.7.1) we obtain $D_l(\lambda_i)\,\boldsymbol{x}_i = \boldsymbol{0}$. Hence the right eigenvectors of A coincide with right latent vectors of $D_l(\lambda)$.

For part (ii) suppose we are given that $\lambda_1, \lambda_2, \ldots, \lambda_n$ are eigenvalues of A and also latent roots of $D_l(\lambda)$, and $\boldsymbol{x}_1, \ldots, \boldsymbol{x}_n$ are right eigenvectors of A and linearly independent right latent vectors of $D_l(\lambda)$. Let

$$X = [\boldsymbol{x}_1, \boldsymbol{x}_2, \ldots, \boldsymbol{x}_n] \quad \text{and} \quad \Lambda = \operatorname{diag}\{\lambda_1, \lambda_2, \ldots, \lambda_n\};$$

then, as in (1.8.5), we have

$$A = X\Lambda X^{-1},$$

and as in (3.4.6) we see that $D_l(A) = 0$.

The proof for a left solvent matrix is similar.

Corollary *All right and left solvents A of $D_l(\lambda)$ satisfy the scalar equation*

$$\Delta(A) = 0.$$

By the Cayley–Hamilton theorem we have $c(A) = 0$; hence (3.7.3) implies that $\Delta(A) = 0$ for all right solvents, and similarly for left solvents.

This provides a generalization of the Cayley–Hamilton theorem, for, if

$$D_1(\lambda) \equiv A - \lambda I,$$

then

$$\Delta(A) = c(A) = 0.$$

Before proceeding we should perhaps note that given the latent roots of a λ-matrix (even a simple λ-matrix) we cannot arbitrarily assign n of these to be the eigenvalues of a solvent matrix. This is because the associated latent vectors must also be eigenvectors of the solvent matrix. Now eigenvectors of a matrix associated with unequal eigenvalues are necessarily linearly independent (Theorem 2.9), whereas this is not necessarily true of latent vectors of a λ-matrix of degree greater than one. Thus, those unequal latent roots which are also eigenvalues of some solvent matrix must have linearly independent latent vectors associated with them.

Finally we prove a generalization of our special case of Sylvester's theorem (see (2.4.10)).

THEOREM 3.8 *Let $D_l(\lambda)$ be a square λ-matrix of order n, and let A be a matrix of order n with simple structure. Denote the eigenvalues of A by $\mu_1, \mu_2, \dots, \mu_n$ and the constituent matrices of A by $G_1, G_2, \dots, G_n$; then*

$$D_l(A) = \sum_{i=1}^{n} D_l(\mu_i)\, G_i, \tag{3.7.4}$$

and

$$\hat{D}_l(A) = \sum_{i=1}^{n} G_i D_l(\mu_i). \tag{3.7.5}$$

Proof We have

$$D_l(A) = A_0 A^l + A_1 A^{l-1} + \dots + A_l,$$

and, from (2.4.8),

$$A^m = \sum_{i=1}^{n} \mu_i^m G_i, \qquad m = 0, 1, 2, \dots, l.$$

Hence,

$$D_l(A) = \sum_{i=1}^{n} (A_0\mu_i^l + A_1\mu_i^{l-1} + \cdots + A_l)\, G_i$$
$$= \sum_{i=1}^{n} D_l(\mu_i)\, G_i.$$

A similar proof gives (3.7.5). If A is a right solvent of $D_l(\lambda)$, then Theorem 3.7 implies that each component $D_l(\mu_i)\, G_i$ must vanish.

CHAPTER 4

LAMBDA-MATRICES, II

4.1 Introduction

In this chapter we will be concerned almost exclusively with simple λ-matrices $D_l(\lambda)$ as defined in § 3.1. The central result is stated in Theorem 4.3, where we obtain $[D_l(\lambda)]^{-1}$ in spectral form. That is, we express $[D_l(\lambda)]^{-1}$ in terms of the latent vectors and the spectrum of latent roots of $D_l(\lambda)$. The method of proof consists in first transforming the problem to that of inverting a matrix of order ln ($l > 1$) whose elements are linear in λ. We then show that this gives rise to the inversion of a simple matrix pencil, which has already been achieved in Theorem 2.4. By comparing the expression obtained there with the explicit form for the inverse we prove the theorem.

In § 4.2 we consider the problem of converting a λ-matrix to a linear form, and in § 4.4 we discuss the conditions on the latent vectors and obtain some results connected with them. We next obtain an alternative expression for our spectral theorem, followed by some further results pertaining to λ-matrices of the second degree.

In § 4.7, we use the matrix pencil associated with $D_l(\lambda)$ to generalize the Rayleigh quotient mentioned in § 2.8, and prove that the properties described in Theorem 2.14 carry over to our generalized form. Finally, we prove a result which will be useful in the discussion of one of the algorithms described in Chapter 5.

4.2 An Associated Matrix Pencil

It is well known, and easily verified, that the scalar polynomial equation

$$d(\lambda) \equiv a_0\lambda^l + a_1\lambda^{l-1} + \cdots + a_{l-1}\lambda + a_l = 0, \qquad (4.2.1)$$

in which $a_0 \neq 0$, can be written in the determinantal form

$$|M - \lambda I| = 0,$$

where

$$M = \begin{bmatrix} \dfrac{-a_1}{a_0} & \dfrac{-a_2}{a_0} & \cdots & \cdots & \cdots & \dfrac{-a_l}{a_0} \\ 1 & 0 & \cdots & \cdots & \cdots & 0 \\ 0 & 1 & \cdots & \cdots & \cdots & 0 \\ \vdots & \vdots & \vdots & \vdots & \vdots & \vdots \\ 0 & 0 & \cdots & 1 & 0 & 0 \\ 0 & 0 & \cdots & 0 & 1 & 0 \end{bmatrix} \tag{4.2.2}$$

M is known as the *companion matrix* of the polynomial $d(\lambda)$. In fact we have

$$d(\lambda) = a_0 \, |M - \lambda I| \, . \tag{4.2.3}$$

Clearly, the zeros of $d(\lambda)$ coincide with the eigenvalues of M, and it is easily verified that a right eigenvector of M associated with the eigenvalue λ_i is $[\lambda_i^{l-1} \; \lambda_i^{l-2} \; \ldots \; \lambda_i \; 1]'$. This property combined with (4.2.3) will be of use to us in the sequel. However, rather than work with the equation

$$(M - \lambda I)\, \boldsymbol{x} = \boldsymbol{0}, \tag{4.2.4}$$

we choose to transform this to an equation retaining the two properties referred to, but which is also symmetric. Thus, we require a non-singular symmetric matrix B of order l with the property that

$$BM = (BM)'.$$

If we can find such a matrix, then, since

$$(BM - \lambda B)\, \boldsymbol{x} = \boldsymbol{0} \qquad \text{implies} \qquad (M - \lambda I)\, \boldsymbol{x} = \boldsymbol{0},$$

the latent roots of the *symmetric* pencil $BM - \lambda B$ will coincide with the zeros of $d(\lambda)$ and the latent vectors of this pencil will coincide with the right eigenvectors of M.

For our present purposes we need only verify that the following matrix has the required properties. We take

$$B = \begin{bmatrix} 0 & 0 & \cdots & & 0 & a_0 \\ 0 & 0 & \cdots & & a_0 & a_1 \\ & & & a_0 & a_1 & a_2 \\ \vdots & \vdots & \vdots & & \vdots & \vdots \\ 0 & 0 & a_0 & & & \vdots \\ 0 & a_0 & a_1 & & & \\ a_0 & a_1 & a_2 & & \cdots & a_{l-1} \end{bmatrix}$$

in which case

$$BM = \begin{bmatrix} 0 & 0 & \cdots & & 0 & a_0 & 0 \\ 0 & 0 & \cdots & & a_0 & a_1 & 0 \\ 0 & 0 & \cdots & a_0 & a_1 & a_2 & 0 \\ \vdots & \vdots & \vdots & & \vdots & \vdots & \vdots \\ 0 & a_0 & \cdots & & . & a_{l-3} & 0 \\ a_0 & a_1 & \cdots & & a_{l-3} & a_{l-2} & 0 \\ 0 & 0 & \cdots & & 0 & 0 & -a_l \end{bmatrix}. \qquad (4.2.5)$$

We should note that, in general, B is not *uniquely* specified by the above conditions. In particular, if $a_l \neq 0$ any product BM^p (for integer p) could play the role of B in (4.2.5). For, as is easily verified, BM^p is then non-singular and symmetric, and so is $(BM^p)M$.

Another interesting transformation to a regular symmetric pencil is obtained in the case $l = 2$ by noting that

$$a_0\lambda^2 + a_1\lambda + a_2 = |A\lambda + C|,$$

where

$$A = \begin{bmatrix} a_0 & 0 \\ 0 & 1 \end{bmatrix}, \qquad C = \begin{bmatrix} a_1 & ia_2^{1/2} \\ ia_2^{1/2} & 0 \end{bmatrix} \qquad (4.2.6)$$

and $i^2 = -1$.

Now consider a regular λ-matrix of the form (3.1.1) and the related sets of homogeneous equations:

$$D_l(\lambda)\,\mathbf{q} = \mathbf{0} \qquad \text{and} \qquad D'l(\lambda)\,\mathbf{r} = \mathbf{0}. \qquad (4.2.7)$$

Comparing $D_l(\lambda)$ with $d(\lambda)$ of (4.2.1) and considering the properties of the symmetric transformation of M just obtained the following theorem suggests itself.

THEOREM 4.1 *If $D_l(\lambda)$ is a regular λ-matrix and if for any latent vector $\mathbf{t}$ of $D_l(\lambda)$ we denote by $\mathbf{t}^{(s)}$ the product $\lambda^s\mathbf{t}$, λ being the corresponding latent root, then the equation $D_l(\lambda)\,\mathbf{q} = \mathbf{0}$ implies and is implied by the partitioned matrix equation:*

$$\left\{ \begin{bmatrix} 0 & 0 & \cdots & & 0 & A_0 \\ 0 & 0 & \cdots & & A_0 & A_1 \\ \vdots & \vdots & & A_0 & A_1 & A_2 \\ 0 & 0 & A_0 & & \vdots & \vdots \\ 0 & A_0 & A_1 & \cdots & & \\ A_0 & A_1 & A_2 & \cdots & \cdots & A_{l-1} \end{bmatrix} \lambda \right. \qquad (4.2.8)$$

$$+\begin{bmatrix} 0 & 0 & \cdots & 0 & -A_0 & 0 \\ 0 & 0 & \cdots & -A_0 & -A_1 & 0 \\ \vdots & \vdots & -A_0 & -A_1 & -A_2 & 0 \\ \vdots & \vdots & \cdot^{\cdot^{\cdot}} & & \vdots & \vdots \\ 0 & -A_0 & & & -A_{l-3} & 0 \\ -A_0 & -A_1 & \cdots & -A_{l-3} & -A_{l-2} & 0 \\ 0 & 0 & \cdots & 0 & 0 & A_l \end{bmatrix}\Bigg\} \begin{bmatrix} \boldsymbol{q}^{(l-1)} \\ \boldsymbol{q}^{(l-2)} \\ \vdots \\ \boldsymbol{q}^{(1)} \\ \boldsymbol{q} \end{bmatrix} = \boldsymbol{0}, \tag{4.2.8}$$

and similarly for the equation $D_l'(\lambda)\,\boldsymbol{r} = \boldsymbol{0}$.

The truth of this theorem is easily verified on evaluating the products of the matrices with the vector while retaining the partioned form, and remembering that A_0 is non-singular.

Denote the two $ln \times ln$ matrices appearing in (4.2.8) by $\mathscr{A}$ and $\mathscr{C}$ respectively. It is clear that because A_0 is non-singular so is $\mathscr{A}$, and so $\mathscr{A}\lambda + \mathscr{C}$ forms a regular pencil of matrices. In certain contexts it may also be useful to note that, if $D_l(\lambda)$ is symmetric, then so are $\mathscr{A}$ and $\mathscr{C}$. If we write $\boldsymbol{q}^+$ for the $ln \times 1$ vector generated by λ and $\boldsymbol{q}$, and $\boldsymbol{r}^+$ for the corresponding vector generated by λ and $\boldsymbol{r}$, then we may write more concisely:

$$(\mathscr{A}\lambda + \mathscr{C})\,\boldsymbol{q}^+ = \boldsymbol{0}, \tag{4.2.9}$$

and

$$(\mathscr{A}'\lambda + \mathscr{C}')\,\boldsymbol{r}^+ = \boldsymbol{0}. \tag{4.2.10}$$

Now, if $D_l(\lambda)$ is also a *simple* λ-matrix, then a typical latent root λ_m of multiplicity α has α linearly independent associated right latent vectors $\boldsymbol{q}_i$. It is clear that the corresponding vectors $\boldsymbol{q}_i^+$ will also be linearly independent. Thus for each latent root the matrix $\mathscr{A}\lambda_m + \mathscr{C}$ has degeneracy equal to the multiplicity of λ_m. It now follows from Theorem 2.10 that $\mathscr{A}\lambda + \mathscr{C}$ is a *simple* matrix pencil. The converse is easily seen to be true. We therefore have:

THEOREM 4.2 *A regular λ-matrix $D_l(\lambda)$ is simple if and only if the associated pencil, $\mathscr{A}\lambda + \mathscr{C}$, is a simple pencil.*

4.3 The Inverse of a Simple λ-Matrix in Spectral Form

We have seen in § 3.1 that a simple λ-matrix of order n and degree l has ln latent roots, if they are counted according to their multiplicities. Let these be $\lambda_1, \lambda_2, \ldots, \lambda_{ln}$ and let $\Lambda = \text{diag}\,\{\lambda_1, \lambda_2, \ldots, \lambda_{ln}\}$. Further, we may associate a set of independent right latent vectors with any latent root, the number of vectors in the set being equal to the multiplicity of the corresponding root. The union of these sets forms a set of ln vectors: $\mathbf{q}_1, \mathbf{q}_2, \ldots, \mathbf{q}_{ln}$, the ordering of the vectors corresponding to that of the λ's. We write

$$Q = [\mathbf{q}_1, \mathbf{q}_2, \ldots, \mathbf{q}_{ln}], \tag{4.3.1}$$

an $n \times ln$ matrix. Similarly, we define

$$R = [\mathbf{r}_1, \mathbf{r}_2, \ldots, \mathbf{r}_{ln}] \tag{4.3.2}$$

in terms of the left latent vectors. With these definitions we have:

THEOREM 4.3 *If $D_l(\lambda)$ is a simple λ-matrix and λ is not equal to a latent root of $D_l(\lambda)$, then Q and R can be defined in such a way that*

$$\lambda^r[D_l(\lambda)]^{-1} = Q\Lambda^r(I\lambda - \Lambda)^{-1} R', \qquad \text{for} \qquad r = 0, 1, 2, \ldots, l-1 \tag{4.3.3}$$

and

$$\lambda^l[D_l(\lambda)]^{-1} = Q\Lambda^l(I\lambda - \Lambda)^{-1} R' + A_0^{-1}. \tag{4.3.4}$$

Proof In equation (4.2.8) there is displayed the lnth order simple matrix pencil which is associated with $D_l(\lambda)$ (Theorem 4.2). The diagonal matrix of the latent roots of this pencil may be assumed to coincide with Λ, and from the definitions of the latent vectors $\mathbf{q}^+$ and $\mathbf{r}^+$ we see that the $ln \times ln$ matrices of its latent vectors may be written:

$$\mathscr{Q} = \begin{bmatrix} Q\Lambda^{l-1} \\ \vdots \\ Q\Lambda \\ Q \end{bmatrix} \quad \text{and} \quad \mathscr{R} = \begin{bmatrix} R\Lambda^{l-1} \\ \vdots \\ R\Lambda \\ R \end{bmatrix}. \tag{4.3.5}$$

We may now apply Theorem 2.4 to the pencil $\mathscr{A}\lambda + \mathscr{C}$ and write

$$(\mathscr{A}\lambda + \mathscr{C})^{-1} = \mathscr{Q}(I\lambda - \Lambda)^{-1} \mathscr{R}'. \tag{4.3.6}$$

From the proof of Theorem 2.2 it is clear that the choice arising in the definitions of $\mathscr{Q}$ and $\mathscr{R}$ mentioned in Theorem 2.4 refers not only to the normalization of the vectors but also to the definition of columns of $\mathscr{Q}$ and $\mathscr{R}$ corresponding to multiple roots. If we take any linear transformation of a set of vectors corresponding to a multiple latent root, the structure of $\mathscr{Q}$ and $\mathscr{R}$ as defined in (4.3.5) is still preserved, so this leads to no complication. Equations (2.2.1) imply that $\mathscr{Q}$ and $\mathscr{R}$ must be chosen so that

$$\mathscr{R}'\mathscr{A}\mathscr{Q} = I_{ln} \quad \text{and} \quad \mathscr{R}'\mathscr{C}\mathscr{Q} = -\Lambda, \tag{4.3.7}$$

where we use I_{ln} for the unit matrix of order ln, reserving I for the unit matrix of order n. For convenience we now write

$$(\mathscr{A}\lambda + \mathscr{C})^{-1} = \mathscr{B}$$

and denote the $n \times n$ partitions of $\mathscr{B}$ by B_{ij}. On substituting from (4.3.5) into (4.3.6) we find that

$$\begin{aligned} B_{ij} &= Q\Lambda^{l-i}(I\lambda - \Lambda)^{-1}\Lambda^{l-j}R' \\ &= Q\Lambda^{2l-i+j}(I\lambda - \Lambda)^{-1}R'. \end{aligned} \tag{4.3.8}$$

Hence B_{ij} depends only on the sum $i + j$, and there are only $2l - 1$ distinct $n \times n$ partitions of $\mathscr{B}$ given by $i + j = 2, 3, \ldots, 2l$.

Now by definition of $\mathscr{B}$ we have

$$(\mathscr{A}\lambda + \mathscr{C})\,\mathscr{B} = I_{ln}.$$

Writing the left-hand side of this equation in partitioned form, we

$$\begin{bmatrix} 0 & 0 & \cdots & 0 & -A_0 & A_0\lambda \\ 0 & 0 & \cdots & -A_0 & A_0\lambda - A_1 & A_1\lambda \\ \vdots & \vdots & \cdot^{\cdot^{\cdot}} & A_0\lambda - A_1 & A_1\lambda - A_2 & A_2\lambda \\ 0 & -A_0 & & & \vdots & \vdots \\ -A_0 & A_0\lambda - A_1 & \cdots & \vdots & A_{l-3}\lambda - A_{l-2} & A_{l-2}\lambda \\ A_0\lambda & A_1\lambda & \cdots & & A_{l-2}\lambda & A_{l-1}\lambda + A_l \end{bmatrix} [B_{ij}] = I_{ln}. \tag{4.3.9}$$

We use this equation to evaluate the partitions B_{il}. The partition of the product on the left appearing in the $1, l$ position gives

$$-A_0B_{l-1,l} + \lambda A_0B_{l,l} = 0,$$

and, since A_0 is non-singular,

$$B_{l-1,l} = \lambda B_{l,l}. \tag{4.3.10}$$

The partition of the product in the $2, l$ position gives

$$-A_0B_{l-2,l} + (A_0\lambda - A_1)\,B_{l-1,l} + A_1\lambda B_{l,l} = 0,$$

and using (4.3.10) we find that

$$B_{l-2,l} = \lambda^2 B_{l,l}.$$

Continuing in this way up to the partition in the $l - 1, l$ position we find that

$$B_{i,l} = \lambda^{l-i}B_{l,l} \qquad \text{for} \qquad i = 1, 2, \ldots, l - 1. \tag{4.3.11}$$

From the partition in the l, l position we obtain finally

$$A_0B_{1,l}\lambda + A_1B_{2,l}\lambda + \cdots + A_{l-2}B_{l-1,l}\lambda + (A_{l-1}\lambda + A_l)\,B_{l,l} = I,$$

and using (4.3.11)

$$(A_0\lambda^l + A_1\lambda^{l-1} + \cdots + A_{l-1}\lambda + A_l)\,B_{l,l} = I. \qquad (4.3.12)$$

From (4.3.8) and (4.3.12) we obtain

$$B_{l,l} = [D_l(\lambda)]^{-1} = Q(I\lambda - \Lambda)^{-1}\,R',$$

and combining this with (4.3.8) and (4.3.11):

$$B_{i,l} = \lambda^{l-i}[D_l(\lambda)]^{-1} = Q\Lambda^{l-i}(I\lambda - \Lambda)^{-1}\,R',$$

for $i = 1, 2, \ldots, l-1$. These two results establish (4.3.3).

To prove (4.3.4) we consider the partition of (4.3.9) in the 1, 1 position. We obtain

$$-A_0B_{l-1,1} + A_0\lambda B_{l,1} = I,$$

or,

$$\lambda B_{l,1} = B_{l-1,1} + A_0^{-1}.$$

But from (4.3.8) we have $B_{l,1} = B_{1,l}$, and using (4.3.11) we have $B_{1,l} = \lambda^{l-1}B_{l,l}$ and $B_{l-1,1} = Q\Lambda^l(I\lambda - \Lambda)^{-1}\,R'$, so that we obtain

$$\lambda^l B_{l,l} = Q\Lambda^l(I\lambda - \Lambda)^{-1}\,R' + A_0^{-1}.$$

Equation (4.3.4) now follows from (4.3.12), and the theorem is proved.

As in (2.3.2), we note that (4.3.3) and (4.3.4) can also be expressed as finite series of partial fractions. Thus, if we write for the outer products of the latent vectors,

$$F_i = \mathbf{q}_i\mathbf{r}_i', \qquad i = 1, 2, \ldots, ln, \qquad (4.3.13)$$

and use the multiplication lemma of § 1.2, then (4.3.3) can be written

$$\lambda^r[D_l(\lambda)]^{-1} = \sum_{i=1}^{ln}\frac{\lambda_i^r F_i}{\lambda - \lambda_i}, \qquad r = 0, 1, 2, \ldots, l-1, \qquad (4.3.14)$$

and (4.3.4) becomes

$$\lambda^l[D_l(\lambda)]^{-1} = \sum_{i=1}^{ln}\frac{\lambda_i^l F_i}{\lambda - \lambda_i} + A_0^{-1}. \qquad (4.3.15)$$

Finally, we note that, because $\mathscr{A}\lambda + \mathscr{C}$ is a simple matrix pencil, the matrices $\mathscr{Q}$ and $\mathscr{R}$ of (4.3.5) must have rank ln, and this implies that the last n rows of each matrix are linearly independent. Hence:

THEOREM 4.4 *If $D_l(\lambda)$ is a simple λ-matrix, then the $n \times ln$ matrices of latent vectors, Q and R, each have rank n.*

In other words, the set of vectors $\boldsymbol{q}_1, \boldsymbol{q}_2, \dots, \boldsymbol{q}_{ln}$ *spans* the n-dimensional vector space over the complex numbers, and similarly for the $\boldsymbol{r}_i$. Each set of vectors is said to be *complete*.

The relations (4.3.14) allow a further extension of our results. Let $P(\lambda)$ be a λ-matrix of any degree and order $m \times n$. Then there exist unique $m \times n$ λ-matrices $Q(\lambda)$, $R(\lambda)$, with the degree of $R(\lambda)$ less than l, such that

$$P(\lambda) = Q(\lambda)\, D_l(\lambda) + R(\lambda) \tag{4.3.16}$$

(cf. § 3.4). Postmultiplying by the inverse of $D_l(\lambda)$ and writing

$$R(\lambda) = \sum_{j=0}^{l-1} R_{l-j}\lambda^j,$$

we obtain

$$P(\lambda)[D_l(\lambda)]^{-1} = Q(\lambda) + \sum_{j=1}^{l-1} R_{l-j}\lambda^j [D_l(\lambda)]^{-1}.$$

Equation (4.3.14) now gives

$$P(\lambda)[D_l(\lambda)]^{-1} = Q(\lambda) + \sum_{j=0}^{l-1} R_{l-j} \sum_{i=1}^{ln} \frac{\lambda_i^j F_i}{\lambda - \lambda_i}$$

$$= Q(\lambda) + \sum_{i=1}^{ln} \frac{R(\lambda_i)\, F_i}{\lambda - \lambda_i}.$$

Or, alternatively,

$$P(\lambda)[D_l(\lambda)]^{-1} = Q(\lambda) + \left(\sum_{j=0}^{l-1} R_{l-j} Q \Lambda^j\right)(I\lambda - \Lambda)^{-1} R'. \tag{4.3.17}$$

Similarly, if $P_1(\lambda)$ is an $n \times m$ λ-matrix, and

$$P_1(\lambda) = D_l(\lambda)\, \hat{Q}(\lambda) + \hat{R}(\lambda),$$

where the degree of $\hat{R}$ is less than l, then

$$[D_l(\lambda)]^{-1} P_1(\lambda) = \hat{Q}(\lambda) + \sum_{i=1}^{ln} \frac{F_i \hat{R}(\lambda_i)}{\lambda - \lambda_i}$$

$$= \hat{Q}(\lambda) + Q(I\lambda - \Lambda)^{-1} \sum_{j=0}^{l-1} \Lambda^j R' \hat{R}_{l-j}.$$

4.4 Properties of the Latent Vectors

We now investigate in what way the latent vectors of $D_l(\lambda)$ must be defined in order that Theorem 4.3 should hold. The proof of that theorem depends on the fact that the latent vectors (4.3.5) of the simple matrix pencil $\mathscr{A}\lambda + \mathscr{C}$ can be defined so that (4.3.7) is true. This was shown to be possible in Theorem 2.4. Now let λ_1 and λ_2 be any two latent roots of $D_l(\lambda)$ with multiplicities α and β respectively. Let R_α be an $n \times \alpha$ matrix of linearly independent left latent vectors of λ_1 and let Q_β be an $n \times \beta$ matrix of linearly independent right latent vectors of λ_2. Then according to (4.3.7) it is always possible to define R_α and Q_β so that

$$[R'_\alpha\lambda_1^{l-1} \cdots R'_\alpha\lambda_1\ R'_\alpha]\,[\mathscr{A}] \begin{bmatrix} Q_\beta\lambda_2^{l-1} \\ \vdots \\ Q_\beta\lambda_2 \\ Q \end{bmatrix} = K, \qquad (4.4.1)$$

where

$$K = \begin{cases} 0 & \text{if} \quad \lambda_1 \neq \lambda_2, \\ I_\alpha & \text{if} \quad \lambda_1 = \lambda_2, \end{cases} \qquad (4.4.2)$$

and the zero matrix here is of order $\alpha \times \beta$.

Substituting for $\mathscr{A}$ in (4.4.1) (see eqns. (4.2.7) and (4.2.8)) and multiplying the partitioned matrices together, it is found that

$$R'_\alpha\{A_0(\lambda_1^{l-1} + \lambda_2\lambda_1^{l-2} + \lambda_2^2\lambda_1^{l-3} + \cdots + \lambda_2^{l-1}) + A_1(\lambda_1^{l-2} + \lambda_2\lambda_1^{l-3} + \cdots + \lambda_2^{l-2}) + \cdots + A_{l-2}(\lambda_1 + \lambda_2) + A_{l-1}\}\,Q_\beta = K. \qquad (4.4.3)$$

If $\lambda_1 \neq \lambda_2$, this equation combined with (4.4.2) gives

$$R'_\alpha\left\{A_0\frac{\lambda_1^l - \lambda_2^l}{\lambda_1 - \lambda_2} + A_1\frac{\lambda_1^{l-1} - \lambda_2^{l-1}}{\lambda_1 - \lambda_2} + \cdots + A_{l-2}\frac{\lambda_1^2 - \lambda_2^2}{\lambda_1 - \lambda_2} + A_{l-1}\right\}Q_\beta = 0,$$

which may be written

$$R'_\alpha\{D_l(\lambda_1) - D_l(\lambda_2)\}\,Q_\beta = 0. \qquad (4.4.4)$$

Obviously, this condition is always satisfied but it seems to be the best generalized biorthogonality condition we can get.

If $\lambda_2 = \lambda_1$, then (4.4.3) and (4.4.2) give

$$R'_\alpha\{A_0\lambda_1^{l-1} + (l-1)\,A_1\lambda_1^{l-2} + \cdots + 2A_{l-2}\lambda_1 + A_{l-1}\}\,Q_\alpha = I_\alpha.$$

Thus,

$$R'_\alpha D_l^{(1)}(\lambda_1)\, Q_\alpha = I_x. \qquad (4.4.5)$$

This is the essential condition on the latent vectors under which Theorem 4.3 is valid. That is, if $\boldsymbol{r}_\mu$, $\boldsymbol{q}_\nu$ are the latent vectors of λ_i $(\mu, \nu = 1, 2, \ldots, \alpha_i)$, then they must form a biorthogonal system with respect to $D_l^{(1)}(\lambda_i)$; in other words

$$\boldsymbol{r}'_\mu D_l^{(1)}(\lambda_i)\, \boldsymbol{q}_\nu = \delta_{\mu\nu}. \qquad (4.4.6)$$

However, we note that even when this condition is satisfied the matrices R_α and Q_α are indeterminate to the extent of an arbitrary non-singular transformation and its inverse (as in Theorem 2.2). We note also that it is always *possible* to satisfy these biorthogonality conditions provided $D_l(\lambda)$, and hence $\mathscr{A}\lambda + \mathscr{C}$, is a *simple* λ-matrix. We formulate our results as follows:

THEOREM 4.5 *Let λ_i be a latent root of multiplicity α_i $(1 \leqq \alpha_i \leqq n)$ of the simple λ-matrix $D_l(\lambda)$; then basis vectors can be found in the subspaces of left and right latent vectors of λ_i which are biorthogonal with respect to $D_l^{(1)}(\lambda_i)$. That is, there exist right latent vectors $\boldsymbol{q}_\nu$ and left latent vectors $\boldsymbol{r}_\mu$ $(\mu, \nu = 1, 2, \ldots, \alpha_i)$ associated with λ_i such that* (4.4.6) *is satisfied.*

We may immediately infer that $D_l^{(1)}(\lambda_i) \neq 0$ if $D_l(\lambda)$ is a simple λ-matrix. As an example of this situation consider $D_2(\lambda) = A_0\lambda^2 + A_2$, where A_0 and A_2 are respectively non-singular and singular. Since $|A_2| = 0$ it is easily seen that $D_2(\lambda)$ has a null latent root. However, $D_2^{(1)}(0) = 0$, so that $D_2(\lambda)$ cannot be a simple λ-matrix.

We can obtain a result similar to that of Theorem 4.5 together with a converse statement by employing Theorem 2.13. First, Theorem 4.2 assures as that $D_l(\lambda)$ is simple if and only if the associated pencil $\mathscr{A}\lambda + \mathscr{C}$ is simple. Then, using Theorem 2.13 and the notation of eqns. (4.2.9) and (4.2.10), we see that the conditions under which the scalar $\boldsymbol{r}^{+\prime}\mathscr{A}\boldsymbol{q}^+$ vanishes are relevant. As in the simplification of (4.4.1) to (4.4.5) it is found that

$$\boldsymbol{r}^{+\prime}\mathscr{A}\boldsymbol{q}^+ = \boldsymbol{r}' D_l^{(1)}(\lambda)\, \boldsymbol{q}, \qquad (4.4.7)$$

and we may now deduce from Theorem 2.13:

THEOREM 4.6 *The regular λ-matrix $D_l(\lambda)$ is defective if and only if there exists a latent root λ with a right latent vector $\boldsymbol{q}$ such that $\boldsymbol{r}' D_l^{(1)}(\lambda)\, \boldsymbol{q} = 0$ for all left latent vectors $\boldsymbol{r}$ of λ.*

We can now deduce further properties of the latent vectors.

THEOREM 4.7 *If the latent vectors of the simple λ-matrix $D_l(\lambda)$ are defined as indicated in Theorem 4.5 then*

$$Q\Lambda^{l-1}R' = A_0^{-1}, \tag{4.4.8}$$

$$Q\Lambda^{\mu}R' = 0, \quad \mu = 0, 1, 2, \ldots, l-2, \tag{4.4.9}$$

and if in addition A_l is non-singular, then

$$Q\Lambda^{-1}R' = -A_l^{-1}. \tag{4.4.10}$$

Proof Putting $r = 0$ in (4.3.3) we have

$$[D_l(\lambda)]^{-1} = Q(I\lambda - \Lambda)^{-1}R'. \tag{4.4.11}$$

Substitute this expression back into (4.3.4) and we obtain

$$\lambda^l Q(I\lambda - \Lambda)^{-1}R' = Q\Lambda^l(I\lambda - \Lambda)^{-1}R' + A_0^{-1}.$$

Thus,

$$Q(I\lambda^l - \Lambda^l)(I\lambda - \Lambda)^{-1}R' = A_0^{-1},$$

and

$$Q(I\lambda^{l-1} + \Lambda\lambda^{l-2} + \cdots + \Lambda^{l-1})R' = A_0^{-1}.$$

Since this is true for a continuous range of values of λ we may equate the coefficients of different powers of λ to zero and obtain eqns. (4.4.8) and (4.4.9).

If A_l is non-singular, then zero is not a latent root of $D_l(\lambda)$, for $|D_l(0)| = |A_l| \neq 0$. In this case we may put $\lambda = 0$ in (4.4.11) and obtain (4.4.10). This completes the proof.

Equations (4.4.8) and (4.4.10) suggest that the matrices $A_0, A_1, \ldots, A_l$ can all be expressed in terms of the latent vectors and latent roots of $D_l(\lambda)$. This is indeed the case, though the explicit forms of these expressions become very involved. We will obtain a recursive method for calculating these matrices.

Consider any λ for which $|\lambda| > \max|\lambda_i|$, and write (4.3.15) in the form

$$\lambda^l[D_l(\lambda)]^{-1} = A_0^{-1} + \frac{1}{\lambda}\sum_{i=1}^{ln}\lambda_i^l F_i(1 - \lambda_i/\lambda)^{-1}$$

$$= A_0^{-1} + \frac{1}{\lambda}\sum_{i=1}^{ln}\lambda_i^l F_i\,\{1 + \lambda_i/\lambda + \lambda_i^2/\lambda^2 + \ldots\};$$

reverting to the matrices of latent vectors:

$$\lambda^l[D_l(\lambda)]^{-1} = A_0^{-1} + \lambda^{-1}Q\Lambda^l R' + \lambda^{-2}Q\Lambda^{l+1}R' + \ldots.$$

(This formula may be of some value in estimating the inverse of $D_l(\lambda)$ for large values of $|\lambda|$.) If we now premultiply both sides of the last equation by $D_l(\lambda)$ we find that

$$\lambda^l I = \{A_0\lambda^l + A_0\lambda^{l-1} + \cdots + A_l\} \times \times \{A_0^{-1} + \lambda_0^{-1}Q\Lambda^l R' + \lambda^{-2}Q\Lambda^{l+1}R' + \cdots\}.$$

If, for convenience, we denote the matrices $Q\Lambda^r R'$ by Γ_r, then the coefficients of $\lambda^{l-1}, \lambda^{l-2}, \ldots, \lambda^0$ give

$$\left.\begin{aligned} &A_1A_0^{-1} + A_0\Gamma_l = 0, \\ &A_2A_0^{-1} + A_1\Gamma_l + A_0\Gamma_{l+1} = 0, \\ &\vdots \\ &A_lA_0^{-1} + A_{l-1}\Gamma_l + \cdots + A_0\Gamma_{2l-1} = 0; \end{aligned}\right\}, \qquad (4.4.12)$$

from which $A_1, A_2, \ldots, A_l$ may be calculated successively.

4.5 The Inverse of a Simple λ-Matrix in Terms of its Adjoint

Let λ_i, $i = 1, 2, \ldots, s$ be the distinct latent roots of the simple λ-matrix $D_l(\lambda)$, so that $s \leq ln$. Suppose that the latent vectors are defined in accordance with Theorem 4.5 and that the latent roots λ_i have multiplicities α_i. Then there are α_i distinct matrices F_i, as defined in (4.3.13), associated with λ_i. Let the sum of these matrices be H_i, a matrix of rank α_i; then putting $r = 0$ in (4.3.14) we may write

$$[D_l(\lambda)]^{-1} = \sum_{i=1}^{s} \frac{H_i}{\lambda - \lambda_i}, \qquad (\lambda \neq \lambda_i). \qquad (4.5.1)$$

Let $F(\lambda)$ be the adjoint of $D_l(\lambda)$ and recall the latent equation:

$$\Delta(\lambda) \equiv |D_l(\lambda)| = 0.$$

Theorem 4.8 *With the above definitions, if λ_i is a latent root of $D_l(\lambda)$ with multiplicity α, then*

$$H_i = \frac{\alpha F^{(\alpha-1)}(\lambda_i)}{\Delta^{(\alpha)}(\lambda_i)}. \qquad (4.5.2)$$

The proof of this theorem is almost the same as that of Theorem 2.8, which is, of course, a special case. We will simply outline the steps of the proof in this more general context. First of all we need a lemma analogous to that for Theorem 2.8 but stating now that the first non-vanishing derivative $\Delta^{(p)}(\lambda_i)$ is $\Delta^{(\alpha)}(\lambda_i)$, and that

$$\Delta^{(\alpha)}(\lambda_i) = a(\alpha!) \,\Pi'\, (\lambda_i - \lambda_r)^{\alpha_r}, \qquad (4.5.3)$$

where Π' denotes the product from $r = 1$ to $r = s$ excluding $r = i$, and $a = |A_0|$. We also have

$$\Delta(\lambda) = a \prod_{j=1}^{s} (\lambda - \lambda_j)^{\alpha_j}. \tag{4.5.4}$$

We start the proof proper by differentiating the equation

$$F(\lambda) = [D_l(\lambda)]^{-1} \Delta(\lambda)$$

p times by Leibniz's rule and employing (4.5.1). Then, if λ_i is a latent root of multiplicity $p+1$ and if we use the properties of (4.5.4) and its derivatives, it is found that

$$F^{(p)}(\lambda_i) = a(p!)\, H_i \, \Pi' \, (\lambda_i - \lambda_r)^{\alpha_r},$$

and (4.5.3) now gives

$$F^{(p)}(\lambda_i) = \frac{p! H_i \Delta^{(p+1)}(\lambda_i)}{(p+1)!}.$$

The theorem now follows on putting $p + 1 = \alpha$.

Equation (4.3.14) can now be expressed in the following (less elegant) way:

$$\lambda^r [D_l(\lambda)]^{-1} = \sum_{i=1}^{s} \frac{\lambda_i^r \alpha_i F^{(\alpha_i - 1)}(\lambda_i)}{\Delta^{(\alpha_i)}(\lambda_i)(\lambda - \lambda_i)}, \qquad r = 0, 1, 2, \ldots, l-1. \tag{4.5.5}$$

In particular, if all the latent roots are simple, then

$$\lambda^r [D_l(\lambda)]^{-1} = \sum_{i=1}^{ln} \frac{\lambda_i^r F(\lambda_i)}{\Delta^{(1)}(\lambda_i)(\lambda - \lambda_i)}. \tag{4.5.6}$$

4.6 Lambda-matrices of the Second Degree

We consider here the special case $l = 2$. Under the conditions of Theorem 3.5 we can invert equation (3.6.1) and rearrange the result to obtain

$$[D_2(\lambda)]^{-1} = Q_1(I\lambda - \Lambda_1)^{-1} Q_1^{-1} A_0^{-1} \{R_2(I\lambda - \Lambda_2)^{-1} R_2^{-1}\}' \tag{4.6.1}$$

provided λ is not equal to a latent root of $D_2(\lambda)$. This expression may be more convenient than the corresponding case of (4.3.3). In particular, it is not necessary to normalize the eigenvectors or to biorthogonalize those corresponding to multiple roots (as specified in Theorem 4.5) in order to use (4.6.1). We observe also that only half of the total of $4n$ left and right latent vectors are employed in (4.6.1), whereas all of these are employed in (4.3.3). For future reference we will now show how the matrices

$$Q_2 = [\mathbf{q}_{n+1}, \mathbf{q}_{n+2}, \dots, \mathbf{q}_{2n}] \quad \text{and} \quad R_1 = [\mathbf{r}_1, \mathbf{r}_2, \dots, \mathbf{r}_n] \tag{4.6.2}$$

can be expressed in terms of Q_1 and R_2.

We define the non-singular matrix

$$T = (R_2' A_0 Q_1)^{-1}. \tag{4.6.3}$$

Then (3.6.1) can be written

$$D_2(\lambda) = (R_2^{-1})' \, (I\lambda - \Lambda_2) \, T^{-1}(I\lambda - \Lambda_1) \, Q_1^{-1}. \tag{4.6.4}$$

Let λ_ν be a latent root of $D_2(\lambda)$ with a right latent vector $\mathbf{q}_\nu$; then

$$(I\lambda_\nu - \Lambda_2) \, T^{-1}(I\lambda_\nu - \Lambda_1) \, Q_1^{-1}\mathbf{q}_\nu = \mathbf{0}. \tag{4.6.5}$$

Now let ν take any of the values $n+1, n+2, \dots, 2n$; then the root λ_ν appears on the leading diagonal of Λ_2 but not on that of Λ_1. Thus, $(I\lambda_\nu - \Lambda_2)$ is singular whereas $(I\lambda_\nu - \Lambda_1)$ is not. If we now write

$$T^{-1}(I\lambda_\nu - \Lambda_1) \, Q_1^{-1}\mathbf{q}_\nu = \mathbf{q}, \tag{4.6.6}$$

then (4.6.5) becomes

$$(I\lambda_\nu - \Lambda_2) \, \mathbf{q} = \mathbf{0}, \tag{4.6.7}$$

and this equation is satisfied by $\mathbf{q} = \mathbf{e}_{\nu-n}$, where $\mathbf{e}_i$ has elements δ_{ij}, $j = 1, 2, \dots, n$. Hence, from (4.6.6) we may write

$$\mathbf{q}_\nu = Q_1(I\lambda_\nu - \Lambda_1)^{-1} \, T\mathbf{e}_{\nu-n}. \tag{4.6.8}$$

If we write

$$\alpha = \nu - n \quad (\alpha = 1, 2, \dots, n) \tag{4.6.9}$$

and $A_{*\alpha}$ for the αth column of A we have

$$\mathbf{q}_\nu = Q_1(I\lambda_\nu - \Lambda_1)^{-1} \, T_{*\alpha}. \tag{4.6.10}$$

We now introduce the matrix

$$\Omega = \left(\frac{t_{ij}}{\lambda_{n+j} - \lambda_i}\right), \qquad i, j = 1, 2, \dots, n, \tag{4.6.11}$$

and (4.6.10) may be written

$$\mathbf{q}_\nu = Q_1 \Omega_{*\alpha},$$

so that

$$\mathbf{q}_\nu = (Q_1\Omega)_{*\alpha}, \qquad \alpha = 1, 2, \dots, n.$$

It is now apparent from the definition (4.6.2) that we may write

$$Q_2 = Q_1\Omega. \tag{4.6.12}$$

Similarly, it may be proved that

$$R_1' = -\Omega R_2'. \tag{4.6.13}$$

From (4.6.10) we deduce that a root contained in Λ_2 and having multiplicity α will have exactly α linearly independent right latent vectors. Since this result is already implicit in the conditions made on the latent roots contained in Λ_1 we have proved:

THEOREM 4.9 *Under the conditions of Theorem* 3.5, $D_2(\lambda)$ *is a simple* λ*-matrix.*

THEOREM 4.10 *Under the conditions of Theorem* 3.5 (*and with the definitions* (4.6.2) *and* (4.6.11)) *we may write*

$$Q_2 = Q_1\Omega \quad \textit{and} \quad R_1' = -\Omega R_2'.$$

By premultiplying the first of these equations by Q_1^{-1} and then substituting for Ω in the second it is found that

$$Q_1R_1' = -Q_2R_2',$$

or, as in equation (4.4.9) with $\mu = 0$, we may write

$$QR' = 0, \tag{4.6.14}$$

where $Q = [Q_1\ Q_2]$, $R = [R_1\ R_2]$.

This result suggests that the choice of Q_2 and R_1 specified in Theorem 4.10 is related in some way to the conditions specified in Theorem 4.5 (in the case $l = 2$).

Finally, we note that, although Q_1 and R_2 are assumed to be non-singular, the same may not be true of Ω, and hence Q_2 and R_1 may be singular, though they will both have the same rank. To illustrate this point and Theorem 4.10, consider the matrix

$$\begin{bmatrix} 2\lambda(\lambda-4) & -\lambda(\lambda+2) \\ -(\lambda-1)(\lambda-4) & (\lambda-1)(\lambda+2) \end{bmatrix},$$

which is easily seen to be a regular λ-matrix of degree two with latent roots -2, 0, 1, 4. In Theorem 3.5 take

$$\Lambda_1 = \begin{bmatrix} 4 & 0 \\ 0 & -2 \end{bmatrix}, \quad \Lambda_2 = \begin{bmatrix} 0 & 0 \\ 0 & 1 \end{bmatrix}.$$

Then it is easily seen that

$$Q_1 = R_2 = \begin{bmatrix} 1 & 0 \\ 0 & 1 \end{bmatrix},$$

whence

$$T^{-1} = \begin{bmatrix} 2 & -1 \\ -1 & 1 \end{bmatrix} \quad \text{and} \quad T = \begin{bmatrix} 1 & 1 \\ 1 & 2 \end{bmatrix}.$$

Equation (4.6.11) then gives

$$\Omega = \begin{bmatrix} -\frac{1}{4} & -\frac{1}{3} \\ \frac{1}{2} & \frac{2}{3} \end{bmatrix},$$

and we see that $|\Omega| = 0$. Equations (4.6.12) and (4.6.13) now give

$$R_1 = -\Omega' \quad \text{and} \quad Q_2 = \Omega,$$

both of which are singular. The validity of these equations can, of course, be verified by direct calculation of the appropriate latent vectors.

4.7 A Generalization of the Rayleigh Quotient

In § 2.8 we introduced the well-known form of Rayleigh's quotient as applied to symmetric matrix pencils, and also the generalization (2.8.3) to unsymmetric problems. We now show how this can be generalized again for application to λ-matrices.

Let $D_l(\lambda)$ be a simple λ-matrix and recall the definitions contained in eqns. (4.2.8) through (4.2.10). Theorem 4.1 and Rayleigh's quotient in the form (2.8.3) now suggest that we investigate

$$R(\boldsymbol{q}, \boldsymbol{r}, \lambda) = -\frac{\boldsymbol{r}^{+\prime}\mathscr{C}\boldsymbol{q}^{+}}{\boldsymbol{r}^{+\prime}\mathscr{A}\boldsymbol{q}^{+}}. \tag{4.7.1}$$

Notice the dependence of R on λ, which arises from the definitions of $\boldsymbol{q}^{+}$ and $\boldsymbol{r}^{+}$. For the numerator of (4.7.1) we have

$$\begin{aligned}\boldsymbol{r}^{+\prime}\mathscr{C}\boldsymbol{q}^{+} &= [\lambda^{l-1}\boldsymbol{r}' \cdots \lambda\boldsymbol{r}'\ \boldsymbol{r}']\,[\mathscr{C}] \begin{bmatrix} \lambda^{l-1}\boldsymbol{q} \\ \vdots \\ \lambda\boldsymbol{q} \\ \boldsymbol{q} \end{bmatrix} \\ &= \boldsymbol{r}'\{-(l-1)\,\lambda^{l}A_0 - (l-2)\,\lambda^{l-1}A_1 - \cdots - \lambda^2 A_{l-2} + A_l\}\,\boldsymbol{q} \\ &= \boldsymbol{r}'\{-\lambda D_l^{(1)}(\lambda) + D_l(\lambda)\}\,\boldsymbol{q}. \end{aligned} \tag{4.7.2}$$

And for the denominator of (4.7.1) we have

$$\begin{aligned} \boldsymbol{r}^{+\prime}\mathscr{A}\boldsymbol{q}^{+} &= \boldsymbol{r}'\{l\lambda^{l-1}A_0 + (l-1)\,\lambda^{l-2}A_1 + \cdots + 2\lambda A_{l-2} + A_{l-1}\}\,\boldsymbol{q}, \\ &= \boldsymbol{r}'D_l^{(1)}(\lambda)\,\boldsymbol{q}. \end{aligned} \tag{4.7.3}$$

Combining (4.7.1, 2, 3) we obtain

$$R(\boldsymbol{q}, \boldsymbol{r}, \lambda) = \lambda - \frac{\boldsymbol{r}'D_l(\lambda)\,\boldsymbol{q}}{\boldsymbol{r}'D_l^{(1)}(\lambda)\,\boldsymbol{q}} \tag{4.7.4}$$

provided that the denominator on the right does not vanish. We consider the function R as defined for all $\boldsymbol{q}$, $\boldsymbol{r}$, and λ, whether latent values or not. The expression (4.7.4) obviously has one of the usual properties of Rayleigh's quotient: if $\boldsymbol{q}_i, \boldsymbol{r}_i, \lambda_i$ are a set of latent values of the *simple* λ-matrix $D_l(\lambda)$, then by Theorem 4.5 we may assume that $\boldsymbol{r}_i'D_l^{(1)}(\lambda_i)\,\boldsymbol{q}_i \neq 0$, and then

$$R(\boldsymbol{q}_i, \boldsymbol{r}_i, \lambda_i) = \lambda_i. \tag{4.7.5}$$

We now investigate whether the stationary property referred to in Theorem 2.14 carries over to our generalized form. However, before proceeding with this we note that in the following discussion we do not use the polynomial dependence of $D_l(\lambda)$ on λ. Thus, the results obtained will apply to any suitably defined matrix function of the scalar λ. We require only that

$$\boldsymbol{r}_i'D_l^{(1)}(\lambda_i)\,\boldsymbol{q}_i \neq 0$$

for the set of latent values $\boldsymbol{q}_i, \boldsymbol{r}_i, \lambda_i$.

In order to establish the stationary property of R we make arbitrary small variations in $\boldsymbol{q}$, $\boldsymbol{r}$, and λ from a set of latent values and show that the resulting change in R is zero to the first order. Thus, we write $\boldsymbol{q} = \boldsymbol{q}_i + \delta\boldsymbol{q}$, $\boldsymbol{r} = \boldsymbol{r}_i + \delta\boldsymbol{r}$, $\lambda = \lambda_i + \delta\lambda$, so that

$$R(\boldsymbol{q}_i, \boldsymbol{r}_i, \lambda_i) + \delta R = \lambda_i + \delta\lambda - \frac{(\boldsymbol{r}_i' + \delta\boldsymbol{r}')\,D_l(\lambda_i + \delta\lambda)(\boldsymbol{q}_i + \delta\boldsymbol{q})}{(\boldsymbol{r}_i' + \delta\boldsymbol{r}')\,D_l^{(1)}(\lambda_i + \delta\lambda)(\boldsymbol{q}_i + \delta\boldsymbol{q})}. \tag{4.7.6}$$

For the numerator of the quotient on the right we have (assuming $D_l(\lambda)$ to be twice differentiable)

$$\begin{aligned} &(\boldsymbol{r}' + \delta\boldsymbol{r}')\,D_l(\lambda_i + \delta\lambda)\,(\boldsymbol{q}_i + \delta\boldsymbol{q}) \\ &\quad \simeq (\boldsymbol{r}_i' + \delta\boldsymbol{r}')\,\{D_l(\lambda_i) + \delta\lambda D_l^{(1)}(\lambda_i)\}\,(\boldsymbol{q}_i + \delta\boldsymbol{q}) \simeq \delta\lambda\boldsymbol{r}_i'D_l^{(1)}(\lambda_i)\,\boldsymbol{q}_i \end{aligned} \tag{4.7.7}$$

to first order in the variations, using (3.1.2). Combining this with (4.7.6) and using (4.7.5) we find that $\delta R = 0$ to first order in the variations; this being true for an arbitrary variation of $\boldsymbol{q}$, $\boldsymbol{r}$, and λ

about any one of the sets of latent values of $D_l(\lambda)$. In particular, we have:

THEOREM 4.11 *If $D_l(\lambda)$ is a simple λ-matrix and the latent vectors are defined as indicated in Theorem 4.5, then the generalization of Rayleigh's quotient, $R(\mathbf{q}, \mathbf{r}, \lambda)$, given by (4.7.4) is such that $R(\mathbf{q}_i, \mathbf{r}_i, \lambda_i) = \lambda_i$ for $i = 1, 2, \ldots, ln$, and R has a stationary value with respect to λ and the components of $\mathbf{q}$ and $\mathbf{r}$ at each of these points.*

4.8 Derivatives of Multiple Eigenvalues

We refer now to Theorems 2.5, 2.6, and 2.7 and consider the implications for λ-matrices. We have observed that, if λ_i is a latent root of the simple λ-matrix $D_l(\lambda)$, then $D_l^{(1)}(\lambda_i) \neq 0$, so that $q = 1$ in Theorem 2.5. If we also assume that the constant matrix $D_l(\lambda_i)$ has simple structure, then, noting theorem 1.3, we deduce from Theorem 2.5 that there exist $n \times \alpha$ matrices X_α, Y_α whose columns span the subspaces of right and left latent vectors of λ_i, respectively, and such that $Y'_\alpha X_\alpha = I_\alpha$. The derivatives of the null eigenvalue of $D_l(\lambda)$ at $\lambda = \lambda_i$ are then given by the eigenvalues of $Y'_a D_l^{(1)}(\lambda_i)\, X_\alpha$.

However, Theorem 4.5 implies that there exist $n \times \alpha$ matrices Q_α, R_α whose columns also span the subspaces of right and left latent vectors of λ_i respectively, and such that

$$R'_\alpha D_l^{(1)}(\lambda_i)\, Q_\alpha = I_\alpha.$$

Now the columns of X_α and Q_α each provide bases for the same subspace, and similarly for Y_α and R_α. Hence there exist non-singular $\alpha \times \alpha$ matrices S and T such that

$$Q_\alpha = X_\alpha S, \qquad R_\alpha = Y_\alpha T.$$

Thus

$$R'_\alpha D_l^{(1)} Q_\alpha = T'(Y'_\alpha\, D_l^{(1)} X_\alpha)\, S = I_\alpha,$$

and

$$Y'_\alpha D_l^{(1)} X_\alpha = (ST')^{-1}.$$

In particular, we deduce that $Y'_\alpha D_l^{(1)}\, X_\alpha$ is non-singular, and therefore has all its eigenvalues non-zero. Thus, even though λ_i may be a multiple latent root of $D_l(\lambda)$, the α eigenvalues $\mu(\lambda)$ of $D_l(\lambda)$ which vanish at $\lambda = \lambda_i$ are such that their derivatives at λ_i are non-zero. Thus we obtain:

THEOREM 4.12 *Let $D_l(\lambda)$ be a simple λ-matrix with latent root λ_i, and let the constant matrix $D_l(\lambda_i)$ have simple structure; then the eigenvalues $\mu(\lambda)$ of $D_l(\lambda)$ which vanish at $\lambda = \lambda_i$ have finite and non-zero derivatives $\mu^{(1)}(\lambda)$ at $\lambda = \lambda_i$.*

Again, as we noted after the statement of Theorem 2.5, the condition that $D_l(\lambda_i)$ have simple structure can be relaxed to some extent. We need only require that the null eigenvalue of $D_l(\lambda_i)$ have its index equal to its multiplicity, or (Theorem 3.2, corollary 3) that the null eigenvalue have only linear elementary divisors in $(I\mu - D_l(\lambda_i))$.

Let us illustrate the theorem in the case where λ_i is an unrepeated latent root. In this case there exist right and left latent vectors $\boldsymbol{q}$, $\boldsymbol{r}$ of λ_i such that

$$\boldsymbol{r}'D_l^{(1)}(\lambda_i)\,\boldsymbol{q} = 1.$$

However, this tells us nothing about the value of $\boldsymbol{r}'\boldsymbol{q}$. But using Theorem 2.12, what we can say is: If $\boldsymbol{r}'\boldsymbol{q} = 0$, then $D_l(\lambda_i)$ is *not* of simple structure and the eigenvalues $\mu(\lambda)$ which approach zero as $\lambda \to \lambda_i$ may not be differentiable at $\lambda = \lambda_i$; but, if $\boldsymbol{r}'\boldsymbol{q} \neq 0$, then the derivative $\mu^{(1)}(\lambda_i)$ exists and is non-zero.

Theorems 2.6 and 2.7 now give sufficient conditions for the null eigenvalues of $D_l(\lambda_i)$ to be twice differentiable at $\lambda = \lambda_i$. In particular, we observe that, if $D_l(\lambda_i)$ *and* $Y_\alpha' D_l^{(1)}(\lambda_i)\,X_\alpha$ are of simple structure, then the eigenvalues of $D_l(\lambda)$ which vanish at $\lambda = \lambda_i$ are twice differentiable at $\lambda = \lambda_i$.

CHAPTER 5

SOME NUMERICAL METHODS FOR LAMBDA-MATRICES

5.1 Introduction

It is well known, and has been demonstrated in § 4.2 that any problem involving a λ-matrix of order n and degree l can be transformed to a corresponding linear problem involving a matrix pencil of order ln. This has meant that, when faced with such a non-linear latent root problem, the usual procedure has been to transform to an associated linear problem before starting any computation, and then to use any of the numerous methods available for this formulation. Another alternative is to compute the coefficients of the latent equation (3.1.3) and then use a standard method for the solution of a polynomial equation. However, this alternative is fraught with difficulties. In particular, the accuracy with which the polynomial coefficients must be calculated in order to produce negligible errors in the roots can be phenomenal. In this connection Wilkinson's works ([1] and [2]) should be consulted.

In this chapter we consider numerical methods which deal directly with the λ-matrix, and so avoid the cumbersome transformations to associated linear forms and also the inherent pitfalls in the evaluation of the latent equation.

We do not claim to be exhaustive, even in this limited sphere. The algorithms we develop are all of a certain genre. They are all iterative processes, that is, processes of successive approximations, and they generally have good local convergence properties, whereas convergence in the large is indeterminate. For the computation of latent roots the principal competitor for our algorithms appears to be Muller's method (Muller, [1]) as employed by Tarnove [1]. However, Muller's method is not directly applicable to the stability problem, for which our algorithms are very useful. This question will be discussed in § 5.9. The reader should perhaps be warned that the algorithms of § 5.2–§ 5.4 do not turn out to be the most useful, and if he is simply seeking a useful algorithm for computing latent roots, then he might turn at once to § 5.5.

Before discussing particular algorithms let us clarify what we mean by the *rate* of convergence of a convergent sequence λ_s, $s = 0, 1, 2, \ldots$, to its limit λ. If r is a real number which exceeds one, we say that convergence is of order r if

$$\frac{|\lambda_{s+1} - \lambda|}{|\lambda_s - \lambda|^r} \to \text{(finite constant)}$$

as $s \to \infty$. Obviously the higher the value of r the more rapid will the convergence be (for sufficiently large s). If $r = 2$ or 3 the rate of convergence is said to be quadrate or cubic, respectively. The convergence is said to be linear if

$$\frac{|\lambda_{s+1} - \lambda|}{|\lambda_s - \lambda|} \to k$$

as $s \to \infty$, and $0 < k < 1$.

We should also note that the λ-matrix methods may be applied when a large order linear problem is transformed to a smaller order non-linear problem. For example, let A be a square matrix of order $2n$ with $n \times n$ partitions A_{ij} $(i, j = 1, 2)$ in an obvious notation, and suppose that we want to find the eigenvalues and eigenvectors of A, i.e., values of a scalar μ and vectors $\boldsymbol{x}$ of order $2n$ for which

$$(A - \mu I_{2n})\,\boldsymbol{x} = \boldsymbol{0}.$$

If $\boldsymbol{x}' = [\boldsymbol{x}_1'\,\boldsymbol{x}_2']$ where $\boldsymbol{x}_1$ and $\boldsymbol{x}_2$ are each of order n, then this equation is equivalent to the pair of equations

$$(A_{11} - \mu I)\,\boldsymbol{x}_1 + A_{12}\boldsymbol{x}_2 = \boldsymbol{0},$$
$$A_{21}\boldsymbol{x}_1 + (A_{22} - \mu I)\,\boldsymbol{x}_2 = \boldsymbol{0}.$$

Suppose now that A_{21} is non-singular so that $\boldsymbol{x}_1$ can be eliminated from these equations. We then find that

$$[A_{21}^{-1}\mu^2 - (A_{21}^{-1}A_{22} + A_{11}A_{21}^{-1})\,\mu + (A_{11}A_{21}^{-1}A_{22} - A_{12})]\,\boldsymbol{x}_2 = 0,$$

and we obtain a regular λ-matrix of degree two and order n whose latent roots coincide with the eigenvalues of the matrix A of order $2n$.

5.2 A Rayleigh Quotient Iterative Process

We assume that we are given an approximation λ_0 to a latent root, λ, of the square matrix D. We construct a sequence λ_s, $s = 0, 1, 2, \ldots$, and two sequences of vectors, $\boldsymbol{\xi}_s$ and $\boldsymbol{\eta}_s$ by the following rules:

Using arbitrary vectors $\boldsymbol{w}$ and $\boldsymbol{z}$, we form

$$\boldsymbol{\xi}_s = [D(\lambda_s)]^{-1}\boldsymbol{w}, \qquad \boldsymbol{\eta}_s = [D'(\lambda_s)]^{-1}\boldsymbol{z}; \tag{5.2.1}$$

and then

$$\begin{aligned} \lambda_{s+1} &= R(\boldsymbol{\xi}_s, \boldsymbol{\eta}_s, \lambda_s) \\ &= \lambda_s - \frac{\boldsymbol{\eta}_s' D_l(\lambda_s)\, \boldsymbol{\xi}_s}{\boldsymbol{\eta}_s' D_l^{(1)}(\lambda_s)\, \boldsymbol{\xi}_s}. \end{aligned} \tag{5.2.2}$$

Roughly speaking, the formulae (5.2.1) are expected to give approximations for latent vectors of the root we seek, and reference to (4.7.4) then suggests that (5.2.2) will give an improved estimate of the latent root. When $n = 1$, the algorithm is simply the Newton–Raphson method for finding the zeros of a polynomial of degree l. We will refer to this Rayleigh quotient method as the *RQ* algorithm.

Iterative processes of this type have been discussed by Ostrowski [2] and by Crandall [1] for some special linear λ-matrices, and they both demonstrate that, provided $|\lambda_0 - \lambda|$ is sufficiently small, the sequence λ_s tends towards the latent root λ, the convergence being *quadratic*. Convergence of this order for the iterative process defined by (5.2.1) and (5.2.2), when applied to *simple* λ-matrices of any degree, and the existence of a convergence neighborhood for each root have been established (Lancaster [3]), but will not be reproduced here. The proof is based on the spectral decomposition of Theorem 4.3. It should be noted that the results are valid for *multiple* latent roots of simple λ-matrices.

Although the vectors $\boldsymbol{w}$ and $\boldsymbol{z}$ appearing in (5.2.1) are described as arbitrary, there is one condition to be satisfied by each of them; the reason for this will be made clear shortly. If the sequence λ_s converges to the latent root λ, then there must exist at least one pair of latent vectors $\boldsymbol{q}$ and $\boldsymbol{r}$ associated with λ such that

$$\boldsymbol{r}'\boldsymbol{w} \neq 0 \text{ and } \boldsymbol{q}'\boldsymbol{z} \neq 0. \tag{5.2.3.}$$

In practice there will be zero probability of either of these conditions being violated, so the use of the word arbitrary may be excused. These inequalities imply that $\boldsymbol{w}$ and $\boldsymbol{z}$ must not be orthogonal to the subspaces of left and right latent vectors of λ respectively.

An obvious difficulty arises if the matrices A_r of (3.1.1) are all real and we seek a latent root which is complex. If $\boldsymbol{w}$, $\boldsymbol{z}$, and λ_0 are all chosen to be real, then the process breaks down, because the operations involved in (5.2.1) and (5.2.2) will all be real, so that the iterates can never converge to a complex set of latent values. In cases of doubt a complex λ_0 should therefore be chosen.

For the remainder of this chapter we adopt the convention of using an ordinary italic subscript for one of a sequence of iterates to a latent root, and a Greek subscript for a typical latent root. For example, λ_s, λ_ν would be an approximation to a root and a latent root respectively.

If it transpires that the sequence λ_s tends towards a latent root λ, which is a simple zero of $|D_l(\lambda)|$, then the sequences of vectors $\boldsymbol{\xi}_s$ and $\boldsymbol{\eta}_s$ defined by (5.2.1) tend towards the (unmormalized) latent vectors $\boldsymbol{q}$ and $\boldsymbol{r}$. This is apparent if we write $[D_l(\lambda_s)]^{-1}$ in the form (4.3.14) (for a simple λ-matrix) when we have:

$$\boldsymbol{\xi}_s = [D_l(\lambda_s)]^{-1}\,\boldsymbol{w} = \sum_{\nu=1}^{ln} \frac{F_\nu \boldsymbol{w}}{\lambda_s - \lambda_\nu} \tag{5.2.4}$$

Now $F_\nu\,\boldsymbol{w} = (\boldsymbol{q}_\nu \boldsymbol{r}'_\nu)\,\boldsymbol{w} = \boldsymbol{q}_\nu(\boldsymbol{r}_\nu \boldsymbol{w}) = \boldsymbol{q}_\nu \omega_\nu$, say, where

$$\omega_\nu = \boldsymbol{r}'_\nu \boldsymbol{w}. \tag{5.2.5}$$

If we denote the latent root to which the λ_s converge by λ, and the corresponding latent vectors by $\boldsymbol{q}$ and $\boldsymbol{r}$, and put $\boldsymbol{r}'\boldsymbol{w} = \omega$, then we may combine (5.2.4) and (5.2.5) to give

$$\boldsymbol{\xi}_s = \frac{\omega}{\lambda_s - \lambda}\boldsymbol{q} + \sum_{\nu=1}^{ln}{}' \frac{\omega_\nu}{\lambda_s - \lambda_\nu}\boldsymbol{q}_\nu, \tag{5.2.6}$$

where $\sum'$ denotes a summation in which the approximated latent root is omitted. Thus, provided λ_s is sufficiently close to λ and the first of the inequalities (5.2.3) is satisfied, the dominant part of $\boldsymbol{\xi}_s$ is proportional to $\boldsymbol{q}$.

Similarly, we obtain (for a simple latent root, λ)

$$\boldsymbol{\eta}_s = \frac{\zeta}{\lambda_s - \lambda}\boldsymbol{r} + \sum_{\nu=1}^{ln}{}' \frac{\zeta_\nu}{\lambda_s - \lambda_\nu}\boldsymbol{r}, \tag{5.2.7}$$

where

$$\zeta_\nu = \boldsymbol{q}'_\nu \boldsymbol{z}. \tag{5.2.8}$$

If the latent root has multiplicity α, and $1 \leq \alpha \leq n$, then there are α latent vectors, $\boldsymbol{q}_\mu$, say, corresponding to λ which are linearly independent (for a *simple* λ-matrix). In this case, the vector $\omega\boldsymbol{q}$ appearing in (5.2.6) must be replaced by $\sum\omega_\mu \boldsymbol{q}_\mu$, and ξ_s will tend towards a linear combination of the α vectors $\boldsymbol{q}_\mu$. By taking α linearly independent vectors $\boldsymbol{w}$, the various limits correspond to α linearly independent combinations of the $\boldsymbol{q}_\mu$ and the subspace of right latent vectors of λ may be completely determined.

A second iterative process known as broken iteration is considered by Ostrowski [2] and Crandall [1], and suggests the following modified process:

For any λ_s, define

$$\xi_s = [D_l(\lambda_s)]^{-1}\,\xi_{s-1}, \qquad \boldsymbol{\eta}_s = [D'(\lambda_s)]^{-1}\,\boldsymbol{\eta}_{s-1} \tag{5.2.9}$$

for $s = 0, 1, 2, \ldots$, where ξ_{-1} and $\boldsymbol{\eta}_{-1}$ are arbitrary starting vectors $\boldsymbol{w}$ and $\boldsymbol{z}$ satisfying (5.2.3), *and put*

$$\lambda_{s+1} = R(\xi_s, \boldsymbol{\eta}_s, \lambda_s). \tag{5.2.10}$$

The arbitrary vectors $\boldsymbol{w}$ and $\boldsymbol{z}$ used at *each step* of the former scheme are now replaced by the successive approximations ξ_{s-1}, $\boldsymbol{\eta}_{s-1}$ to latent vectors $\boldsymbol{q}$ and $\boldsymbol{r}$. When $l = 1$ it can be shown that, if the above scheme is slightly modified to take advantage of the biorthogonality properties of the latent vectors, then *cubic* convergence can be achieved (see Ostrowski [2]). However, it appears that if $l > 1$ this rate of convergence generally cannot be achieved. Using broken iteration may increase the rate of convergence, but probably not by another order of magnitude.

5.3 Numerical Example for the *RQ* Algorithm

This example is taken from the work of Frazer *et al.* [1], p. 328, and arose from a study of the oscillations of a wing in an airstream. With one or two obvious modifications to the numerical values the problem reduces to finding the latent roots and vectors of $D_2(\lambda)$, where

$$A_0 = \begin{bmatrix} 17.6 & 1.28 & 2.89 \\ 1.28 & 0.824 & 0.413 \\ 2.89 & 0.413 & 0.725 \end{bmatrix}, \qquad A_1 = \begin{bmatrix} 7.66 & 2.45 & 2.10 \\ 0.230 & 1.04 & 0.223 \\ 0.600 & 0.756 & 0.658 \end{bmatrix},$$

$$A_2 = \begin{bmatrix} 121 & 18.9 & 15.9 \\ 0 & 2.70 & 0.145 \\ 11.9 & 3.64 & 15.5 \end{bmatrix}.$$

From experience of similar problems a person experienced in aeroelastic work might take as first approximations to the largest latent root and associated right latent vector:

$$\lambda_0 = i(10.0) \qquad \text{and} \qquad \boldsymbol{w}' = [0 \quad 0 \quad 1.0].$$

For convenience we take $\boldsymbol{z} = \boldsymbol{w}$ in eqns. (5.2.1). After inverting the matrix $D_2(\lambda_0)$ we find from (5.2.1) that

$$\xi_0 = \begin{bmatrix} -0.1481 + i(0.00289) \\ -0.2764 + i(0.01700) \\ 1.0 \end{bmatrix}, \qquad \eta_0 = \begin{bmatrix} -0.1476 + i(0.00528) \\ -0.2716 + i(0.02126) \\ 1.0 \end{bmatrix},$$

where, to facilitate comparisons, we normalize the vectors so that the largest element is one. Substituting λ_0, ξ_0, and η_0 in (5.2.2) (and using the fact that $\eta_s' D_l(\lambda_s)\, \xi_s = \boldsymbol{w}'\xi_s$) it is found that

$$\lambda_1 = -0.756 + i(8.535).$$

This completes one cycle of the iteration. In this case the choice of $\boldsymbol{w}$ is not decisive in determining the question of convergence. It is found that, with the above choice of λ_0 and with $\boldsymbol{w}' = [1.0 \ \ 0 \ \ 0]$ or $\boldsymbol{w}' = [0 \quad 1.0 \quad 0]$ the process would still converge, and hence would do so for an arbitrary choice of $\boldsymbol{w}$.

In the next cycle of operations it is found that

$$\xi_1 = \begin{bmatrix} -0.14938 + i(0.004416) \\ -0.27756 - i(0.020362) \\ 1.0 \end{bmatrix},$$

$$\eta_1 = \begin{bmatrix} -0.14663 - i(0.005382) \\ -0.26753 - i(0.023272) \\ 1.0 \end{bmatrix},$$

and then

$$\lambda_2 = -0.8834 + i(8.4411).$$

To this accuracy the latent root is in fact

$$-0.8848 + i(8.4415).$$

The value given by Frazer *et al.* is $-0.885 + i(8.443)$, so the fourth significant figure of the calculations quoted there is suspect. Bearing this in mind, the corresponding right latent vector as given in the original work is (after renormalization)

$$\begin{bmatrix} -0.1494 + i(0.00466) \\ -0.2778 - i(0.02066) \\ 1.0 \end{bmatrix}$$

We observe that in this case results which are sufficiently accurate for most engineering purposes are obtained after only two steps of the iteration. A further important feature of this and succeeding methods is the possibility of finding latent roots in any order. This is in contrast to the iterative (power) method used by Frazer and his collaborators which gives the dominant root first. The accuracy with which subdominant roots are calculated then depends on the accuracy of those already found.

However, the example used here may not be typical of problems met in practice, as all the latent roots are well separated in the complex λ-plane. The complete set of latent roots is, to three decimal places:

$$-0.885 \pm i(8.442), \qquad 0.097 \pm i(2.522), \qquad -0.920 \pm i(1.760).$$

5.4 The Newton–Raphson Method

Let $D_l(\lambda)$ be a simple λ-matrix with a latent root λ_ν of multiplicity α, and suppose that the null eigenvalue of $D_l(\lambda_\nu)$ has only linear elementary divisors. Let $\mu(\lambda)$ be one of the α eigenvalues of $D_l(\lambda)$ for which $\mu(\lambda_\nu) = 0$ (cf. § 4.8), and let $\boldsymbol{x}(\lambda)$, $\boldsymbol{y}(\lambda)$ be right and left eigenvectors associated with $\mu(\lambda)$. Theorem 2.5 implies that there exist eigenvectors $\boldsymbol{x}(\lambda)$, $\boldsymbol{y}(\lambda)$ and a neighborhood of λ_ν within which $\boldsymbol{y}'\boldsymbol{x} = 1$ and

$$\mu^{(1)}(\lambda) = \boldsymbol{y}'(\lambda)\, D^{(1)}(\lambda)\, \boldsymbol{x}(\lambda). \tag{5.4.1}$$

Premultiplying $D\boldsymbol{x} = \mu\boldsymbol{x}$ by $\boldsymbol{y}'$, we also obtain

$$\mu(\lambda) = \boldsymbol{y}'(\lambda)\, D(\lambda)\, \boldsymbol{x}(\lambda). \tag{5.4.2}$$

We have proved in Theorem 4.12 that $\mu^{(1)}(\lambda_\nu) \neq 0$, and, if $\mu^{(2)}(\lambda_\nu)$ exists also (Theorems 2.6, 2.7,) then it can be proved that there exists a neighborhood $N(\lambda_\nu)$ of λ_ν in which the Newton–Raphson process, when applied to $\mu(\lambda)$, yields a sequence of iterates λ_s, $s = 1, 2, \ldots$, which converges to λ_ν, and that the convergence is monotonic in $N(\lambda_\nu)$ and quadratic as $\lambda_s \to \lambda_\nu$ (Chapter 7 of Ostrowski [3]). Of course we do not justify the algorithm every time we use it, but our discussion shows that there is, in general, a high probability for the existence of $N(\lambda_\nu)$, even though λ_ν may be a multiple latent root. Our NR algorithm is therefore:

Given an approximation λ_0 to a latent root we form the sequence

$$\lambda_{s+1} = \lambda_s - \frac{\mu(\lambda_s)}{\mu^{(1)}(\lambda_s)}, \qquad s = 1, 2, \ldots,$$

$$= \lambda_s - \frac{\boldsymbol{y}_s' D_l(\lambda_s)\, \boldsymbol{x}(\lambda_s)}{\boldsymbol{y}_s' D_l^{(1)}(\lambda_s)\, \boldsymbol{x}(\lambda_s)}. \tag{5.4.3}$$

This equation is obtained from (5.4.1) and (5.4.2).

Notice that it is necessary to calculate at least one eigenvalue $\mu(\lambda_s)$ and a pair of eigenvectors $\boldsymbol{x}(\lambda_s)$, $\boldsymbol{y}(\lambda_s)$ at *each step* of the iteration, so that the non-linear latent root problem is replaced by a sequence of linear eigenvalue problems. This fact immediately condemns the algorithm from a computational point of view, for the time taken in eigenvalue calculations would generally be prohibitive. However, the process will suggest more useful routines in the stability problem to be discussed.

If the practical application of the algorithm were pursued, a strategy for selecting the appropriate eigenvalue $\mu(\lambda_s)$ would have to be determined, and it would probably be found necessary to compute all the eigenvalues at each step. This was in fact done by Howland [1] in his computations with linear problems. If this is the case, then all the results obtained at each step would be utilized by using a refinement of the Newton–Raphson process which employs the second derivative at μ at λ_s. For example, if $\mu_s^{(i)}$ denotes $\mu^{(i)}(\lambda_s)$, $i = 0, 1, 2$, then the algorithm

$$\lambda_{s+1} = \lambda_s - \frac{\mu_s \mu_s^{(1)}}{(\mu_s^{(1)})^2 - \frac{1}{2}\mu_s \mu_s^{(2)}}, \qquad s = 0, 1, 2, \ldots, \tag{5.4.4}$$

provides a sequence which will give cubic convergence as $\lambda_s \to \lambda_\nu$, a simple zero of the regular function $\mu(\lambda)$. The second derivative would, of course, be computed by means of Theorems 2.6 and 2.7.

5.5 Methods Using the Trace Theorem

The next algorithm we consider is elementary in principle; we simply apply the Newton–Raphson process to find the zeros of the polynomial $\Delta(\lambda) = |D(\lambda)|$. Thus we employ the iterative formula:

$$\lambda_{s+1} = \lambda_s - \frac{\Delta(\lambda_s)}{\Delta^{(1)}(\lambda_s)}, \qquad s = 0, 1, 2, \ldots. \tag{5.5.1}$$

The unusual feature of the formula lies in the fact that the coefficients of $\Delta(\lambda)$ need not be calculated explicitly. We employ the following result, referred to above as "the trace theorem."

THEOREM 5.1 *If the elements of the square matrix $D(\lambda)$ are differentiable functions of λ, then, for any λ for which $\Delta(\lambda) \neq 0$, we have*

$$\Delta^{(1)}(\lambda) = \Delta(\lambda)\, Tr\{D^{-1}(\lambda)\, D^{(1)}(\lambda)\} = Tr\{(\text{adj } D(\lambda))\, D^{(1)}(\lambda)\}. \tag{5.5.2}$$

In order to preserve some continuity in our discussion of algorithms we defer the proof of the theorem to the appendix to this chapter. We should, however, make it clear that the notation $Tr(A)$, where A is an $n \times n$ matrix, denotes *the trace of* A and is the sum of the elements on the leading diagonal of A. Thus

$$Tr(A) = \sum_{i=1}^{n} a_{ii}.$$

If we define

$$f(\lambda) = \frac{\Delta^{(1)}(\lambda)}{\Delta(\lambda)} = Tr\{D^{-1}(\lambda)\, D^{(1)}(\lambda)\}, \tag{5.5.3}$$

then the algorithm (5.5.1) may be written:

$$\lambda_{s+1} = \lambda_s - \frac{1}{f(\lambda_s)}, \qquad s = 0, 1, 2, \ldots, \tag{5.5.4}$$

and will be referred to as the *NT* algorithm. The success of the algorithm now depends on a good strategy for determining λ_0, and on the availability of an accurate matrix inversion process to be applied at each step. We note that in computing $f(\lambda_s)$ we need not calculate the entire product $D^{-1}(\lambda_s)\, D^{(1)}(\lambda_s)$, but only the n leading diagonal elements of this product.

Algorithms which utilize the second derivative of $\Delta(\lambda)$ in order to obtain higher rates of convergence are also feasible. We will illustrate with three such algorithms. Suppose now that the elements of $D(\lambda)$ are twice differentiable. Differentiating (5.5.3) it is found that

$$f^{(1)}(\lambda) = \frac{d}{d\lambda}\left(\frac{\Delta^{(1)}(\lambda)}{\Delta(\lambda)}\right) = Tr\left\{\left(\frac{d}{d\lambda} D^{-1}(\lambda)\right) D^{(1)}(\lambda)\right\} + Tr\{D^{-1}(\lambda)\, D^{(2)}(\lambda)\}.$$

But if we differentiate both sides of the equation $D(\lambda)\, D^{-1}(\lambda) = I$ with respect to λ it is found that

$$\frac{d}{d\lambda} D^{-1} = -D^{-1} D^{(1)} D^{-1};$$

whence

$$f^{(1)}(\lambda) = Tr\{D^{-1}D^{(2)} - (D^{-1}D^{(1)})^2\}. \qquad (5.5.5)$$

In order to compute $f^{(1)}(\lambda)$ we therefore use the inversion required to evaluate $f(\lambda)$, and then the further labor is involved mainly in computing the entire product $D^{-1}D^{(1)}$. If we need $\Delta^{(2)}/\Delta$ explicitly it is easily seen that

$$\frac{\Delta^{(2)}}{\Delta} = f^{(1)} + f^2. \qquad (5.5.6)$$

A refinement of the Newton–Raphson process giving cubic convergence to simple roots (used in (5.4.4)) is then

$$\lambda_{s+1} = \lambda_s - \frac{2f(\lambda_s)}{f^2(\lambda_s) - f^{(1)}(\lambda_s)}, \qquad s = 0, 1, 2, \ldots. \qquad (5.5.7)$$

We will refer to this as algorithm $NT1$.

An alternative refinement giving quadratic convergence to all roots, whatever their multiplicities, is obtained by applying the classical Newton–Raphson process to the function $\Delta(\lambda)/\Delta^{(1)}(\lambda)$, which has only simple zeros. This gives rise to algorithm $NT2$:

$$\lambda_{s+1} = \lambda_s + \frac{f(\lambda_s)}{f^{(1)}(\lambda_s)}, \qquad s = 0, 1, 2, \ldots. \qquad (5.5.8)$$

Finally we note the Laguerre method, which is known to have very useful convergence properties (Todd [1], Durand [1], Parlett [1]). Locally, the algorithm gives cubic convergence to unrepeated roots, and linear convergence to multiple roots. If all the roots are *real*, then this algorithm will give a sequence converging to a root *whatever the initial guess*, λ_0, *may be*. This property does not carry over to the case of complex roots, but even in this case the properties of convergence in the large seem to compare favourably with those of our preceding algorithms. The recursion formula is $\lambda_{s+1} = g(\lambda_s)$, where

$$g(\lambda) = \lambda - \frac{k\Delta}{\Delta^{(1)} \pm \sqrt{(k-1)\{(k-1)(\Delta^{(1)})^2 + k\,\Delta\Delta^{(2)}\}}},$$

and k is a constant. In terms of $f(\lambda)$ and $f^{(1)}(\lambda)$, (eqns. (5.5.3) and (5.5.5)) we have

$$g(\lambda) = \lambda - \frac{k}{f(\lambda) \pm \sqrt{(k-1)\{-kf^{(1)}(\lambda) - f^2(\lambda)\}}}. \qquad (5.5.9)$$

The sign is determined in the denominator of the fraction on the right so as to maximize the modulus of this denominator, that is, in such a way that the smaller correction to the preceeding iterate is chosen. When applied to a polynomial, the constant k is usually assigned to be the degree of the polynomial. However, the rates of convergence claimed are preserved if k is any number greater than the degree of the polynomial. Thus, in computing with an $n \times n$ λ-matrix of degree l we assign $k = ln$. This is the degree of $\Delta(\lambda)$ if the λ-matrix is regular, but the same rates of convergence are attained even though the matrix is not regular. We refer to this as the *LT* algorithm.

This remark brings us to an important feature of all the algorithms described in this section, namely, that the rates of convergence ascribed do not depend on the nature of the elementary divisors of the root we are attempting to compute. In particular, we need not assume that $D(\lambda)$ is a simple λ-matrix. This may be contrasted with the Rayleigh quotient method and the Newton–Raphson method of § 5.4 in which the rates of convergence claimed for multiple latent roots require that their elementary divisors be linear.

Another advantage of the algorithms of this section is the facility with which roots already computed can be "divided out" so that subsequent sequences of iterates cannot converge to a root already found. Suppose that at some stage of a calculation the latent roots $\lambda_1, \lambda_2, \ldots, \lambda_s$ have been computed, and let

$$p(\lambda) = \prod_{j=1}^{s} (\lambda - \lambda_j).$$

Then the algorithms are reformulated so that in computing the $(s + 1)^{st}$ root we replace the polynomial $\Delta(\lambda)$ by $\Delta(\lambda)/p(\lambda)$. It can be verified that this is achieved if we replace $f(\lambda)$ and $f^{(1)}(\lambda)$ in (5.5.4), (5.5.7), (5.5.8), and (5.5.9) by

$$\left.\begin{aligned} & f(\lambda) - \sum_{j=1}^{s} \frac{1}{\lambda - \lambda_j} \\ \text{and} \qquad & f^{(1)}(\lambda) + \sum_{j=1}^{s} \frac{1}{(\lambda - \lambda_j)^2} \end{aligned}\right\} \tag{5.5.10}$$

respectively. The sums on the right must then be computed at each step of the iteration. However, the time involved in these computations is small compared to that required for the inversion.

5.6 Iteration of Rational Functions

If a good approximation to a latent root is known, then this is generally used for λ_0 and, if this approximation is good enough, the sequence of iterates resulting from one of the above algorithms will converge to the root. However, we suggest that our algorithms NT, $NT1$, $NT2$, and LT will generally be effective, even though little or nothing is known about the distribution of the latent roots in the complex plane. A more or less arbitrary strategy for selecting the first approximation, λ_0, can be used though in some cases (see Parlett [1], for instance) there is a great deal to be gained by a careful determination of this strategy.

Nevertheless, before attempting iterations from arbitrary starting values, and for the purpose of comparing our algorithms, it is well to consider some general aspects of the theory of iteration. This theory originates largely with Julia [1] and Fatou [1].

We denote by $\varphi(z)$ the function with which we iterate in the complex z-plane. Thus, our algorithms are to be written in the form $z_{s+1} = \varphi(z_s)$, $s = 0, 1, 2, \ldots$. For the purposes of this discussion we assume $\varphi(z)$ to be a rational function of degree k. That is, $\varphi(z)$ is the ratio of two polynomial functions of z having no zeros in common, and k is the larger of the degrees of these polynomials. Our discussion is then immediately applicable to the algorithms RQ, NT, $NT1$, and $NT2$, and it may be assumed that phenomena similar to those we describe will arise in the application of NR and LT.

Given a number z_0, we define the sequence of *consequents* (iterates) of z_0 by $z_{r+1} = \varphi(z_r)$, $r = 0, 1, 2, \ldots$. We may also enquire into the points which are mapped onto z_0 by φ, that is, the solutions of the z-equation, $z_0 = \varphi(z)$. It is easily seen that there are k such points given by the k branches of the inverse function φ^{-1}. These points are called *antecedents* of z_0. Given some means of selecting a particular branch of φ^{-1}, we can then define a unique sequence of antecedents of any z_0.

A point corresponding to the number z for which $z = \varphi(z)$ is called a *fixed point* of the iteration. It is easily seen that in the four algorithms covered by this discussion the latent roots of $D(\lambda)$ are fixed points. We then subdivide the fixed points as follows: If ζ is a fixed point of $\varphi(z)$ and $|\varphi'(\zeta)| < 1$, then ζ is a *point of attraction.* If the fixed point is such that $|\varphi'(\zeta)| > 1$, then ζ is a *point of repulsion.* The significance of these definitions is apparent in the following results:

If ζ is a point of attraction (repulsion) then there exists a neighborhood† $\mathcal{N}$ *of ζ such that, if z_0 is in $\mathcal{N}$, then the sequence of consequents z_ν of z_0 is such that $z_\nu \to \zeta (z_\nu \nrightarrow \zeta)$ as $\nu \to \infty$.*

In the case of a point of repulsion, $z_\nu \nrightarrow \zeta$ implies that the sequence of consequents does not have ζ as a limit, though we must exclude the possibility that $z_\nu = \zeta$ for some finite ν. It is easily see that, if $0 < |\varphi'(\zeta)| < 1$, then a sequence z_ν with z_0 in $\mathcal{N}$ converges linearly to ζ. If $\varphi'(\zeta) = 0$, then the rate of convergence is some number $r > 1$ (cf. § 5.1).

It can be verified that the latent roots of $D(\lambda)$ are points of attraction for the iteration functions appropriate to any of the RQ, NT, $NT1$, or $NT2$ algorithms, and the above result ensures the existence of a convergence neighborhood for each of them. Furthermore, the latent roots of $D(\lambda)$ are the only points of attraction. The questions now arise: What is the nature of the set of all points which are antecedents of points in the convergence neighborhood of a given root? Can the whole plane be divided into a finite number of regions such that all points in any one region give rise to a sequence of consequents having a particular root as its limit?

Unfortunately analysis shows that the geometry of these domains of convergence is generally so complicated that no clear answer can be given for the first question, and the answer to the second question is, for all practical purposes, "no." Indeed, the only case known to the writer in which the answer is affirmative is that of the Newton–Raphson iteration applied to a quadratic equation! (See appendix F of Ostrowski [3]).

Analysis of the domains of convergence is found to depend critically on the following concept—which is also of importance for the practitioner in numerical analysis. We define an *n-cycle* to be a set of n distinct points $\zeta, \zeta_1, \ldots, \zeta_{n-1}$, say, such that

$$\zeta_1 = \varphi(\zeta), \quad \zeta_2 = \varphi(\zeta_1), \ldots, \quad \zeta = \varphi(\zeta_{n-1}), \qquad n \geqq 2.$$

If we introduce the notation $\varphi_2(z) = \varphi(\varphi(z))$, $\varphi_3(z) = \varphi(\varphi_2(z))$, etc., then ζ is such that $\zeta = \varphi_n(\zeta)$. Thus a point belonging to an n-cycle

† A "neighborhood of ζ" implies the set of points in a circular disc of the complex plane with center ζ.

of φ is a fixed point of φ_n. It should be noted that

$$\varphi_n'(\zeta) = \varphi'(\zeta)\,\varphi'(\zeta_1)\,\ldots\,\varphi'(\zeta_{n-1}) = \varphi_n'(\zeta_i)$$

for $i = 1, 2, \ldots, n-1$. There will therefore be no ambiguity if we define a convergent (divergent) n-cycle to be one for which $|\varphi_n'(\zeta)| < 1$ (> 1), where ζ is a point of the n-cycle.

An example of a convergent 2-cycle is obtained if we apply the Newton–Raphson process to the polynomial $z^3 - 3z^2 + z - 1$. It is found that

$$\varphi(z) = \frac{2z^3 - 3z^2 + 1}{3z^2 - 6z + 1},$$

which has the 2-cycle, $\zeta = 0$, $\zeta_1 = 1$. Furthermore, either of these points is a point of attraction of φ_2 for which convergent sequences of consequents attain quadratic convergence.

We now observe that, as well as having regions of the plane associated with points of attraction, there will generally be finite regions associated with convergent n-cycles and, if we are unfortunate enough to start an iteration in one of these, then the sequence of consequents will converge in a cyclic manner to the points of the n-cycle. An alarming prospect is now revealed. The function $\varphi_n(z)$ is rational of degree k^n. It appears that φ_n may therefore have as many as k^n points of attraction and this for every positive integer n. There may even be infinitely many convergent n-cycles with $n \geqq 2$, each having its finite domain of convergence. This situation may not leave much space for the domains of convergence of the points of attraction of φ. Fortunately, however, the number of convergent n-cycles is finite, and hope is rekindled by the following results of Julia:

If the number of points of attraction of φ is n_1, and the number of convergent n-cycles of φ is n_2, then

$$n_1 + n_2 \leqq 2(k-1), \qquad n_1 \leqq k. \tag{5.6.1}$$

In the example given above we have $k = 3$, $n_1 = 3$, and hence $n_2 = 1$. Thus, the convergent 2-cycle illustrated is the *only* convergent n-cycle for all $n \geqq 2$.

From the point of view of practical computation the divergent n-cycles are generally not troublesome, as there is only a very small probability of ever hitting on a member of such a cycle. The convergent n-cycles are more troublesome, and do occur in practice, though possibly with a lower frequency than we deserve.

Algorithms in which they may arise should include some means of detecting such a state of affairs and beginning the iteration again with a new starting value.

The inequalities (5.6.1) suggest that n_2 may increase with k. Now the RQ algorithm will generally have $k = (2n - 1)\,l$, and if $\Delta(\lambda)$ has degree d, ($d = ln$ when $D(\lambda)$ is regular), the algorithms NT, $NT1$, and $NT2$ will generally have $k = d, 2d - 1$, and $2d - 1$ respectively. The suggestion therefore arises that convergent n-cycles are more likely to arise in the algorithms $NT1$, $NT2$, and RQ than in NT. This seems to be borne out by experience and presents one argument in favor of the straightforward NT method.

5.7 Behavior at Infinity

There is always some danger that an iterate will approach a singularity of the iterating function φ very closely. The effect of this will be to throw the next iterate a great distance from the origin in the complex plane. This distance may be very large compared to the modulus of the latent root furthest from the origin. We may now ask: Will succeeding iterates return to the neighborhood of the roots, and if so, in how many steps? We may take it that the origin is included in this neighborhood.

In order to investigate this question we examine the behavior of $\varphi(z)$ as $z \to \infty$. This is easily done for our algorithms by simply retaining the dominant terms of the polynominals involved as $z \to \infty$. If we write $\varphi(\lambda) \sim \varphi_1(\lambda)$ as $\lambda \to \infty$, the values for φ_1 displayed in the table are obtained. In these results σ_1 and σ_2 denote the sum and the sum of the squares of the latent roots respectively.

Algorithm	RQ	NT	$NT1$	$NT2$
$\varphi_1(\lambda)$	$\left(\frac{l-1}{l}\right)\lambda$	$\left(\frac{d-1}{d}\right)\lambda$	$\left(\frac{d-1}{d+1}\right)\lambda$	$\frac{\sigma_1}{d}$

For the Laguerre algorithm the behavior when $k = d$ is quite different from that when $k \neq d$. It is found that

$$\varphi_1(\lambda) = \frac{1}{d}\left\{\sigma_1 \pm \sqrt{(d-1)(d\sigma_2 - \sigma_1^2)}\right\}, \qquad \text{for} \qquad k = d,$$

$$\varphi_1(\lambda) = \frac{\sqrt{\beta}}{d + \sqrt{\beta}}\,\lambda, \qquad \text{for} \qquad k > d,$$

where $\beta = d(k-1)(k-d)$.

For algorithms RQ, NT, $NT1$ and LT with $k > d$, the iterates of a point λ with $|\lambda|$ very large will lie close to the radial line from the origin to λ; their magnitudes will decrease at each step in the ratio dictated by the appropriate entry in the table. These ratios are generally too big for this risk to be run—it may take many steps to return to the neighborhood of the roots after an iterate has approached a singularity of φ. This situation can be remedied by checking the magnitude of each iterate. If one becomes too big, then restart the computation with a new λ_0. This possibility can safely be ignored in the algorithms $NT2$, and LT with $k = d$, for an iterate of very large modulus is immediately followed by an iterate in the neighborhood of the roots.

5.8 A Comparison of Algorithms

In this section we give an account of some numerical experiments using the algorithms NT, $NT1$, $NT2$, and LT. The experiments were designed to give some insight into the problem of resolving clustered roots. It is well known that, when roots are clustered, or coincide, in the complex plane, the accuracy with which they can be determined (using a fixed computer word length) usually decreases. Certain errors, often of an oscillatory nature, arise and are not improved no matter how many steps of the iteration are taken. It was therefore thought desirable to construct a test problem in which the matrix coefficients depend on a parameter whose value determines the proximity of clustered roots. The following λ-matrix has this property. We consider $A_0\lambda^2 + A_1\lambda + A_0$, where $A_0 = I$,

$$A_1 = \begin{bmatrix} 3\alpha & -(1+\alpha^2+2\beta^2) & \alpha(1+2\beta^2) & -\beta^2(\alpha^2+\beta^2) \\ 2 & 0 & 0 & 0 \\ 0 & 2 & 0 & 0 \\ 0 & 0 & 2 & 0 \end{bmatrix},$$

$$A_2 = \begin{bmatrix} -1+2\alpha^2 & \alpha-\alpha(\alpha^2+2\beta^2) & 2\alpha^2\beta^2 & -\alpha\beta^2(\alpha^2+\beta^2) \\ 2\alpha & -(\alpha^2+2\beta^2) & 2\alpha\beta^2 & -\beta^2(\alpha^2+\beta^2) \\ 1 & 0 & 0 & 0 \\ 0 & 1 & 0 & 0 \end{bmatrix},$$

and $\beta = 1 + \alpha$. The distribution of the eight latent roots is illustrated in Fig. 5.1. The roots at the origin and $\pm i$ do not depend on α.

When $\alpha = 0$ there are triple roots at $\pm i$ and a double root at the origin. The problem was devised to represent the distribution of latent roots in a rather small, but badly conditioned stability problem of the kind considered later in this chapter.

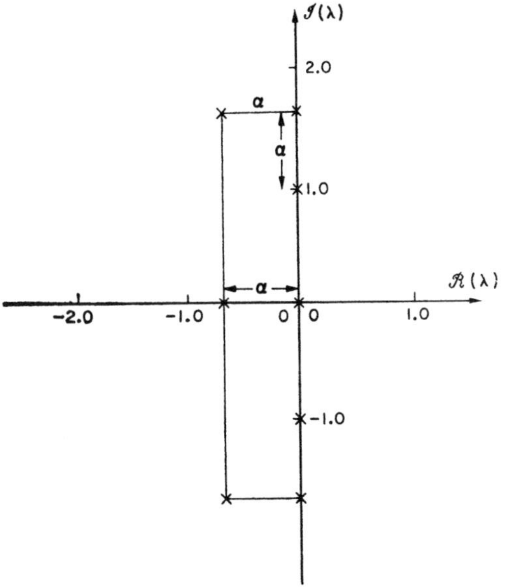

FIG. 5.1. Latent roots for the test problem.

A sequence of iterations was stopped when either

$$\left.\begin{array}{lll} & |\lambda_s - \lambda_{s-1}|/|\lambda_s| < \varepsilon & \text{if } \max\{R(\lambda_s),\ I(\lambda_s)\} > \varepsilon \\ \text{or} & |\lambda_{s+1} - \lambda_s| < \varepsilon & \text{otherwise,} \end{array}\right\} \qquad (5.8.1)$$

where ε was a negative integral power of 10 and as small as possible. If higher exponents were used than those tabulated for ε below, the irremovable errors referred to above would prevent the determination of the roots by this criterion. The arithmetic was performed in floating-point decimal form with a ten digit mantissa. Values of α were used which permitted exact determination of the matrix coefficients. In every case the Gauss–Jordan matrix inversion process was used (Rutishauser [1]). Although this is not the fastest possible process, it was generally found to be very accurate. The results obtained with algorithms NT, $NT1$, $NT2$, and LT

at $\alpha = 0.5$, 0.1, 0.02, and $\alpha = 0$ are tabulated below.† The same strategy was used for each algorithm with regard to starting values. The decreasing accuracy as α decreases is apparent in two forms. First, the least permissible value of ε increases markedly, so that the number of determinate significant figures in the latent roots decreases accordingly. Second, as the rate of convergence decreases, the maximum error in the computed roots tends to increase relative to ε.

		Total number of cycles				Maximum error/ε			
α	ε	NT	$NT1$	$NT2$	LT	NT	$NT1$	$NT2$	LT
0.5	10^{-8}	43	32	53	27	0.3	0.2	0.2	0.3
0.1	10^{-8}	61	42	65	37	0.9	1	1	0.9
0.02	10^{-7}	62	45	66	36	0.4	0.7	5	0.4
0	10^{-6}	149	65	84	74	35	36	60	44

It might have been expected that the algorithm $NT2$ would be the most successful in the resolution of clustered roots, since this is the only algorithm giving quadratic (or better) convergence to multiple roots. However, our example suggests that this is not the case. Indeed, the algorithms $NT1$ (eqn. (5.5.7)) and LT (eqn. 5.5.9)) are undoubtedly the most successful in every case. The algorithm LT could probably be further improved if, instead of using the strategy which we consistently employed for iterating in the large, we were to take full advantage of the strategy outlined by Parlett [1], which is peculiar to the Laguerre method.

5.9 Algorithms for a Stability Problem

Consider an $n \times n$ matrix D whose elements are differentiable functions of the parameters v and λ, and consider those v, λ for

† To obtain an independent check on the value of these algorithms the four cases tabulated were also solved using Muller's method (cf. § 5.1). This is basically an iterative interpolation process based on efficient evaluation of Δ (λ). It gives convergence of order 1.7, approximately, to simple roots and linear convergence to multiple roots. The time taken per cycle was considerably less than that in any of the above algorithms, but the much larger number of iterations required resulted in a total time of computation comparable to that for the LT method. However, the slower rate of convergence gives rise to larger errors. These were at an acceptable level in cases α — 0.5, 0.1 and 0.02, but were unacceptable (20×10^4 in the above table) at $\alpha = 0$.

which $\Delta(v, \lambda) = |D(v, \lambda)| = 0$. Suppose that v is chosen as the independent variable so that λ may be interpreted as a latent root of $D(\lambda)$ dependent on v. We write $\Delta\{v, \lambda(v)\} = 0$ and consider the behavior of the curves $\lambda(v)$ in the complex λ-plane. We are to investigate the intersections of such a curve with the pure imaginary axis, and find the critical values of v for which this occurs together with the corresponding imaginary part of λ.

The description of this as a stability problem stems from the theory of ordinary differential equations (as developed in the next chapter) where λ arises in a factor $e^{\lambda t}$ of the solution and t is the time. Clearly, such a solution is stable or unstable according as $\mathscr{R}(\lambda) < 0$ or $\mathscr{R}(\lambda) > 0$; the critical cases arise when there is a latent root λ such that $\mathscr{R}(\lambda) = 0$.

If we denote partial derivatives by suffixes, the equation $\Delta\{v, \lambda(v)\} = 0$ implies that

$$\frac{d\lambda}{dv} = -\frac{\Delta v}{\Delta_\lambda},$$

assuming that $\lambda(v)$ is differentiable and $\Delta_\lambda \neq 0$. We now confine our attention to cases in which v is a *real* parameter, and write $\lambda = \mu + i\omega$, where μ, ω are real. The above equation then implies that

$$\frac{d\mu}{dv} = -\mathscr{R}\left(\frac{\Delta_v}{\Delta_\lambda}\right).$$

Now Theorem 5.1 (the trace theorem) implies that

$$\Delta_v = \Delta \, Tr(D^{-1}D_v) \qquad \text{and} \qquad \Delta_\lambda = \Delta \, Tr(D^{-1}D_\lambda),$$

and hence

$$\frac{d\mu}{dv} = -\mathscr{R}\left\{\frac{Tr\,(D^{-1}D_v)}{Tr\,(D^{-1}D_\lambda)}\right\}. \tag{5.9.1}$$

Let us recall that our aim is to find a v such that $\mu(v) = 0$ for some latent root $\lambda(v)$. The last equation now suggests a method for doing this—provided an approximation to the solution is already known. We first assign $v = v_0$, and compute λ_0, the latent root of $D(v_0, \lambda)$, which is expected to go unstable. Then compute $(d\mu/dv)$ at v_0, λ_0 using (5.9.1). The Newton–Raphson algorithm then suggests that a better approximation to the critical value of v is given by

$$v_1 = v_0 - \frac{\mu(v_0)}{\mu^{(1)}(v_0)}. \tag{5.9.2}$$

We may then repeat the process indefinitely.

This suggestion is impractical for computational purposes because an exact latent root is required at each step of the calculation, and this could require an exorbitant amount of effort. However, a practical algorithm can be retrieved if we replace the exact calculation of a latent root at v_0 by one step (or possibly more) of the algorithm NT (eqn. (5.5.4)) towards the latent root, from an approximation λ_0 to this root. Let λ_1 be the result of this step, and let $\lambda_1 = \mu_1 + i\omega_1$; then estimate $d\mu/dv$ by evaluating the right-hand side of (5.9.1) at v_0, λ_1, and then v_1 is given by (5.9.2) using our computed approximations for μ and $d\mu/dv$.

Thus, starting with initial estimates v_0, λ_0 *we define sequences by*

$$\lambda_{s+1} = \lambda_s - \frac{1}{\{Tr\,(D^{-1}D_\lambda)\}_{s,s}}, \tag{5.9.3}$$

$$v_{s+1} = v_s - \frac{\mu_{s+1}}{\mathscr{R}\left\{\dfrac{Tr(D^{-1}D_v)}{Tr(D^{-1}D_\lambda)}\right\}_{s,s+1}}, \tag{5.9.4}$$

where $\lambda_s = i_s + i\omega_s$ *and suffixes on the parentheses indicate the suffixes for* v, λ *respectively at which the matrices are to be evaluated.*

In each cycle of the algorithm two inversions are required: first of $D(v_s, \lambda_s)$ and then of $D(v_s, \lambda_{s+1})$. We will therefore denote this algorithm by $STB2$.

Unlike the algorithms in preceding sections we do *not* recommend iterations in the large for this problem. The algorithm must be viewed as a method of improving known approximations to the critical values. For example, all the latent roots could be found for some values of v and an estimate of a critical condition found by interpolation. This estimate may then be improved by the methods of this section.

The algorithm $STB2$ appears to be inefficient in that two inversions are required at each step, and the matrices involved may not be very different. The algorithm $STB1$ is therefore suggested. This is identical with the above except that the suffixes s, $s + 1$ of (5.9.4) are replaced by s, s. Slower rates of convergence and smaller domains of convergence can be expected for $STB1$ than for $STB2$. Note that the domain of convergence to a particular critical value can be viewed as a volume in the μ, ω, v space.

A third algorithm for the stability problem is obtained from a second method of calculating the derivative $d\mu/dv$ of (5.9.1). For any latent root $\lambda(v)$ there exists a right latent vector $\boldsymbol{q}(v)$ such that

$$D\{v, \lambda(v)\}\,\boldsymbol{q}(v) = \boldsymbol{0}.$$

Denoting partial derivatives by suffixes again and assuming differentiability we find that

$$D_v\boldsymbol{q} + D_\lambda\,(d\lambda/dv)\,\boldsymbol{q} + D\boldsymbol{q}_v = \boldsymbol{0},$$

and premultiplying by a left latent vector $\boldsymbol{r}(v)$ of λ, we obtain

$$\frac{d\lambda}{dv} = -\frac{\boldsymbol{r}'D_v\boldsymbol{q}}{\boldsymbol{r}'D_\lambda\boldsymbol{q}},$$

provided the denominator does not vanish. For real v we now have

$$\frac{d\mu}{dv} = -\mathcal{R}\left(\frac{\boldsymbol{r}'D_v\boldsymbol{q}}{\boldsymbol{r}'D_\lambda\boldsymbol{q}}\right). \tag{5.9.5}$$

Using this formula we can now employ the algorithm (5.9.2) successively to improve an estimate v_0 of the critical value of v. Once more, however, we are faced with a latent root calculation at each step. In addition we would now need corresponding right and left latent vectors $\boldsymbol{q}$ and $\boldsymbol{r}$. But again, a feasible algorithm is obtained if these latent values are replaced by good approximations. In this case improved estimates are obtained by use of the Rayleigh quotient broken iteration process of eqns. (5.2.9) and (5.2.10). The greater part of the computation at each step is contained in the computation of the inverse of $D(v_s, \lambda_s)$. The algorithm $STB3$ is now as follows:

Given initial estimates v_0, λ_0 *and arbitrary starting vectors* $\boldsymbol{\xi}_{-1}$, $\boldsymbol{\eta}_{-1}$ *we define*

$$\boldsymbol{\xi}_s = (D^{-1})_{s,s}\,\boldsymbol{\xi}_{s-1}, \qquad \boldsymbol{\eta}'_s = \boldsymbol{\eta}'_{s-1}(D^{-1})_{s,s},$$

and

$$\lambda_{s+1} = \lambda_s - \frac{\boldsymbol{\eta}'_s(D)_{s,s}\,\boldsymbol{\xi}_s}{\boldsymbol{\eta}'_s(D_\lambda)_{s,s}\,\boldsymbol{\xi}_s},$$

$$v_{s+1} = v_s + \frac{\mu_{s+1}}{\mathcal{R}\left\{\dfrac{\boldsymbol{\eta}'_s(D_v)_{s,s}\,\boldsymbol{\xi}_s}{\boldsymbol{\eta}'_s(D_\lambda)_{s,s}\,\boldsymbol{\xi}_s}\right\}},$$

where $\lambda_s = \mu_s + i\omega_s$, *and* $s = 0, 1, 2, \ldots$.

Clearly $\boldsymbol{\xi}_{-1}$, $\boldsymbol{\eta}_{-1}$ should be chosen to approximate the right and left latent vectors of the critical latent root, if this is possible. It can be expected that the domain of convergence would increase as the vectors $\boldsymbol{\xi}_{-1}$, $\boldsymbol{\eta}_{-1}$ approach these latent vectors.

5.10. Illustration of the Stability Algorithms

We apply the three stability algorithms to the matrix

$$D(v,\lambda) = A\lambda^2 + (Bv + D)\lambda + (Cv^2 + E),$$

where

$$A = \begin{bmatrix} 0.950500 & -0.001002 & 0 & 0 \\ -0.008560 & 0.302120 & 0 & 0 \\ 0 & 0 & 1 & 0 \\ 0 & 0 & 0 & 1 \end{bmatrix},$$

$$B = \begin{bmatrix} 0.06150 & -0.03783 & 0 & 0 \\ -0.03323 & 0.20380 & 0 & 0 \\ 0 & 0 & 0 & 0 \\ 0 & 0 & 0 & 0 \end{bmatrix},$$

$$D = \begin{bmatrix} 0 & 0 & 0 & 0 \\ 0 & 0.05636 & 0 & 0 \\ 0 & 0 & 0.03030 & 0 \\ 0 & 0 & 0 & 0.05263 \end{bmatrix},$$

$$C = \begin{bmatrix} -0.03100 & -0.48460 & 0 & 0 \\ 0.16168 & 1.00000 & 0 & 0 \\ 0 & 0 & 0 & 0 \\ 0 & 0 & 0 & 0 \end{bmatrix},$$

$$E = \begin{bmatrix} 0.74940 & -0.00196 & -0.00100 & 0 \\ -0.01680 & 0.22545 & 0.18040 & 0 \\ -0.00980 & 0.26303 & 0.45280 & 0.27270 \\ 0 & 0 & -0.47370 & 0.53289 \end{bmatrix}$$

All the computations are performed in floating-point arithmetic with a ten decimal digit mantissa. Iterations were started in every case with $\mathscr{R}(\lambda_0) = \mu_0 = 0$. In Fig. 5.2 we sketch regions in the v, ω plane with the property that any point in the interior used to

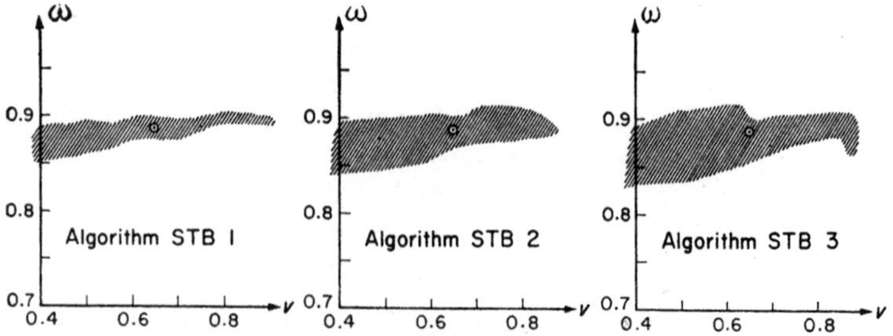

FIG. 5.2. Domains of convergence for the stability algorithms.

give starting values $v_0 = v$, $\lambda_0 = i\omega$ will give sequences converging to the critical condition $v = 0.6475355 \ldots$, $\lambda = i(0.8876455 \ldots)$. For the algorithm *STB*3 the starting vectors

$$\xi'_{-1} = \eta'_{-1} = [1, \quad 1, \quad 1, \quad 1]$$

were used throughout. These regions were obtained by numerical experiments and are not very accurately defined. However, they are sufficiently accurate to indicate the general nature of the results. The criteria by which these algorithms are judged are:

(i) The amount of computation in each cycle.
(ii) The extent of the domain of convergence.
(iii) The rate of convergence of the sequence $v_0, v_1, v_2, \ldots$.

Judged by criterion (i) *STB*1 and *STB*3 are comparable and are both much better than *STB*2. This is because this last algorithm requires two inversions per cycle, whereas the first two require only one. In the example used here, *STB*3 is the best algorithm judged by criterion (ii). As anticipated, *STB*1 is less successful than *STB*2 in this respect. It should be remarked that in each case the domains of convergence suggest that a good approximation to the critical value of λ is required, although a relatively crude estimate of the critical v will suffice. We give in Table 5.1 the iterates obtained in each algorithm using starting values $v_0 = 0.6$ and $\lambda_0 = 0.88$. These results suggest that *STB*3 is again the most advantageous of our algorithms.

TABLE 5.1 SEQUENCES OF ITERATES FOR THE STABILITY ALGORITHMS
STB 1

s	λ_s		v_s
	Real	Imaginary	
0	0	0.88	0.6
1	−0.00132 84748	0.88505 43565	0.66233 31345
2	−0.00000 85261	0.88848 11638	0.66275 87945
3	0.00024 41403	0.88837 40013	0.64739 82793
4	−0.00001 37686	0.88764 41492	0.64827 62417
5	0.00001 23252	0.88768 08616	0.64753 92633
6	0.00000 00358	0.88764 57597	0.64753 71132
7	0.00000 00263	0.88764 56459	0.64753 55377
8	0	0.88764 55709	0.64753 55374
9			
10			

STB 2

s	λ_s		v_s
	Real	Imaginary	
0	0	0.88	0.6
1	−0.00126 52659	0.88493 36226	0.65402 98306
2	−0.00010 33847	0.88796 95051	0.66024 61338
3	0.00019 82700	0.88824 76615	0.64743 85285
4	−0.00000 91903	0.88766 11471	0.64798 88815
5	0.00000 71971	0.88766 68651	0.64755 61499
6	0.00000 03432	0.88764 71200	0.64753 55676
7	−0.00000 00035	0.88764 56139	0.64753 57765
8	0.00000 00035	0.88764 55831	0.64753 55658
9	0.00000 00005	0.88764 55726	0.64753 55382
10	0	0.88764 55710	0.64753 55373

STB 3

s	λ_s		v_s
	Real	Imaginary	
0	0	0.88	0.6
1	−0.00063 13457	0.88560 48255	0.62563 12260
2	−0.00039 05618	0.88661 54970	0.64596 27286
3	−0.00002 56889	0.88757 15158	0.64745 08452
4	−0.00000 14125	0.88764 15457	0.64753 51509
5	−0.00000 00072	0.88764 55503	0.64753 55376
6	0	0.88764 55709	0.64753 55386
7			
8			
9			
10			

Thus, the more general deductions made from (i) and the deductions made from (ii) and (iii) for this test problem lead us to recommend *STB*3 as the most useful of our three algorithms.

Appendix to Chapter 5

This appendix contains a proof of Theorem 5.1; the trace theorem of § 5.5. The reader is referred to page 83 for the statement of the theorem.

Let c_{ij} be the cofactor of the element d_{ij} in $|D(\lambda)|$, and let

$$\mathbf{c}_i' = [c_{i1},\ c_{i2},\ \ldots,\ c_{in}], \qquad i = 1, 2, \ldots, n.$$

Writing D_{*j} for the jth column of D we have

$$\mathbf{c}_i' D_{*j} = \delta_{ij}\,\Delta, \qquad i, j = 1, 2, \ldots, n,$$

$(\Delta(\lambda) = |D(\lambda)|)$, or, equivalently

$$\mathbf{c}_i' D = \Delta \mathbf{e}_i', \tag{A.1}$$

where $\mathbf{e}_i$ has a 1 for its ith element and zeros elsewhere.

We also have

$$\Delta^{(1)} = \sum_{i=1}^{n} \Delta_i, \tag{A.2}$$

where Δ_i is the determinant whose ith column is $D_{*i}^{(1)}$ and the remaining columns are those of D. Expanding Δ_i by the ith column we have

$$\Delta_i = \mathbf{c}_i' D_{*i}^{(1)}. \tag{A.3}$$

Now $DD^{-1} = I$ implies that, provided $\Delta \neq 0$,

$$D^{(1)} = -D(D^{-1})^{(1)}\,D, \tag{A.4}$$

and hence

$$D_{*i}^{(1)} = -D\{(D^{-1})^{(1)}\,D\}_{*i}.$$

Substituting these expressions in (A.3) and using (A.1) We find that

$$\Delta_i = -\Delta \mathbf{e}_i'\{(D^{-1})^{(1)}\,D\}_{*i}.$$

We then find from (A.2) that

$$\Delta^{(1)} = -\Delta\ Tr\ \{(D^{-1})^{(1)}\,D\}.$$

But by (A.4) we may write $(D^{-1})^{(1)}\,D = -D^{-1}D^{(1)}$, and the result (5.5.2) now follows.

CHAPTER 6

ORDINARY DIFFERENTIAL EQUATIONS WITH CONSTANT COEFFICIENTS

6.1 Introduction

We must first draw attention to three distinct uses of the word "order" which will arise in this chapter. We have already defined the order of a square matrix and of a column or row vector. In the theory of differential equations the maximal number of differential operators acting successively on a dependent variable and appearing among the terms of the equations is known as the order of the differential equations. All of these usages are so well established that any attempt to change them would lead to worse confusion. We therefore accept them and trust that the context will make the meaning clear in every case.

After developing some preliminary results concerning systems of first order differential equations, we consider problems in which there are n differential equations of order l involving n dependent variables and one independent variable, t, i.e., we consider *ordinary* differential equations. In § 6.2 we formulate a general solution in terms of the classical complementary function and particular integral. In § 6.3 we investigate the special form of the particular integral obtained when the terms depending on t only are exponential functions of t. This will be required for later applications.

In § 6.4 we formulate explicit forms for the arbitrary constants of our general solution in the case where one-point boundary conditions are given, and in § 6.5 we show how the results obtained earlier by formal manipulations of the differential operator can be obtained by means of the Laplace transform method. Finally, we discuss some special results obtained for second order differential equations which will be useful in our later discussion of vibration problems.

In the following work we employ the differential operator $\mathscr{D} = d/dt$, and we will have to interpret expressions of the form $(\mathscr{D} - \lambda)^{-1} f(t)$, where λ is a constant. We define this expression as a function $x(t)$, say, which satisfies the equation

$$(\mathscr{D} - \lambda)\, x(t) = f(t). \tag{6.1.1}$$

Premultiplying the equation by $e^{-\lambda t}$, we see that

$$\mathscr{D}(e^{-\lambda t}\, x(t)) = e^{-\lambda t} f(t),$$

so that, on integration, we may write

$$x(t) = e^{\lambda t} \int_{t_0}^{t} e^{-\lambda \tau} f(\tau)\, d\tau, \tag{6.1.2}$$

where t_0 is an arbitrary constant, and the contribution to $x(t)$ from this *lower* limit of integration is a multiple of the complementary function of (6.1.1). Accordingly, if it is more convenient, we may use the indefinite integral in (6.1.2) and employ the particular integral

$$x(t) = e^{\lambda t} \int^{t} e^{-\lambda \tau} f(\tau)\, d\tau. \tag{6.1.3}$$

6.2 General Solutions

Consider the set of n first order differential equations in the n variables $x_1(t)$, $x_2(t)$, ..., $x_n(t)$ represented by

$$\dot{\boldsymbol{x}} = A\boldsymbol{x},$$

where A is an arbitrary $n \times n$ matrix of constants, and the dot denotes differentiation with respect to t. That is, $\mathscr{D}\boldsymbol{x} = \dot{\boldsymbol{x}}$. For practical applications, note that A may be real or complex. The next three theorems are stated with the hypothesis that A is constant, but it should be noted that not much more analysis is required to prove corresponding results on the hypothesis that the elements of A are continuous functions of t.

Theorem 6.1 *Let A be a constant matrix; then for all finite values of t there exists a unique solution vector $\boldsymbol{x}(t)$ of the differential equation $\dot{\boldsymbol{x}} = A\boldsymbol{x}$ with the boundary condition $\boldsymbol{x}(0) = \mathbf{c}$.*

Notice that there are two important statements here, first that of *existence*, and second that of *uniqueness*. This result is very important in the development of the theory of differential equations and may also be described as well known. The reader is therefore referred to Coddington and Levinson [1] or Bellman [1] for a proof.

The next problem is to obtain explicit solutions to the differential equation and to characterize the set of all solutions of $\dot{\boldsymbol{x}} = A\boldsymbol{x}$ without imposing the boundary condition. We first obtain some explicit solutions of the form $\xi e^{\mu t}$, where ξ and μ are constant. If there is such a solution, then we must have

A is simple

$$(A - \mu I)\,\xi = \boldsymbol{0}.$$

This demonstrates the fact that, if μ is any eigenvalue of A with an associated eigenvector ξ, then the vector

$$\boldsymbol{x}(t) = \xi e^{\mu t}$$

is a solution of $\dot{\boldsymbol{x}} = A\boldsymbol{x}$. We now ask: Are all possible solutions of this type, or must we look for solutions of a more general kind in order to characterize the set of all solutions? Before we can begin to answer this question we must first extend the idea of linear dependence to vector functions.

The vector functions $\boldsymbol{y}_1(t), \boldsymbol{y}_2(t), \ldots, \boldsymbol{y}_p(t)$ are said to be linearly dependent on an interval $a < t < b$ if there exist constants $a_1, a_2, \ldots, a_p$, not all zero, such that

$$a_1\boldsymbol{y}_1(t) + a_2\boldsymbol{y}_2(t) + \cdots + a_p\boldsymbol{y}_p(t) = \boldsymbol{0}$$

for all t in this interval. The vectors $\boldsymbol{y}_1(t), \ldots, \boldsymbol{y}_p(t)$ are linearly independent if this relation implies that $a_1 = a_2 = \cdots = a_p = 0$. In our applications the a's will be real or complex according as the elements of $\boldsymbol{y}_1, \ldots, \boldsymbol{y}_p$ are real or complex valued.

With this definition we may build up the concepts of vector spaces, dimension, basis, etc., in terms of vector functions in a manner analogous to that outlined in § 1.4. We may now state:

THEOREM 6.2 *Let A be a constant matrix of order n; then the set of all solutions of $\dot{\boldsymbol{x}} = A\boldsymbol{x}$ is an n-dimensional vector space.*

This statement implies that there exist solution vectors $\xi_i(t)$, $i = 1, 2, \ldots, n$, which are linearly independent (in the above sense) and are such that *any* solution vector $\boldsymbol{y}(t)$ is a linear combination of the $\xi_i(t)$. If it transpires that there exist real-valued basis vectors ξ_i, then the vector space will be defined over the real numbers, and, if the ξ_i are necessarily complex valued, then the vector space is defined over the field of complex numbers.

Proof In order to show that the set of all solutions is a vector space, we have only to show that, if $\boldsymbol{h}_1, \ldots, \boldsymbol{h}_p$ are a set of solutions, then *all* linear combinations of them are also solutions. Now any linear combination may be written

$$\boldsymbol{x}(t) = \sum_{i=1}^{p} \alpha_i \boldsymbol{h}_i(t)$$

for some set of constants $\alpha_1, \alpha_2, \ldots, \alpha_p$. Then

$$\dot{\boldsymbol{x}} = \sum_{i=1}^{p} \alpha_i \dot{\boldsymbol{h}}_i = \sum_{i=1}^{p} \alpha_i A \boldsymbol{h}_i,$$

since the $\boldsymbol{h}_i$ are such that $\dot{\boldsymbol{h}}_i = A\boldsymbol{h}_i$; and then we have

$$\dot{\boldsymbol{x}} = A \sum_{i=1}^{p} \alpha_i \boldsymbol{h}_i = A\boldsymbol{x},$$

so that the arbitrary linear combination $\boldsymbol{x}$ is also a solution.

It now remains to determine the dimension of the space. We first take a set of basis vectors for $\mathscr{C}_n$ (or possibly $\mathscr{R}_n$). Let these be $\boldsymbol{b}_1, \boldsymbol{b}_2, \ldots, \boldsymbol{b}_n$. Then consider the set of vector functions $\boldsymbol{\xi}_1(t), \ldots, \boldsymbol{\xi}_n(t)$, which, using Theorem 6.1, are the unique solutions of

$$\dot{\boldsymbol{x}} = A\boldsymbol{x}, \qquad \text{with} \qquad \boldsymbol{x}(0) = \boldsymbol{b}_i, \quad i = 1, 2, \ldots n.$$

In order to complete the proof we show that the $\boldsymbol{\xi}_i(t)$ are linearly independent, and that any solution of $\dot{\boldsymbol{x}} = A\boldsymbol{x}$ may be expressed as a linear combination of them. That is, we show that these vectors form a basis for the space of solution vectors, and the dimension of the solution space is then the number of vectors in the basis.

Suppose that the $\boldsymbol{\xi}_i$ are linearly dependent. Then there exist constants $c_1, c_2, \ldots, c_n$ not all zero such that

$$\sum_{i=1}^{n} c_i \boldsymbol{\xi}_i(t) = \boldsymbol{0}$$

for all finite values of t. In particular, putting $t = 0$, we have

$$\sum_{i=1}^{n} c_i b_i = \boldsymbol{0},$$

which contradicts the assumption that the $\boldsymbol{b}_i$ are independent. Hence the solution vectors $\boldsymbol{\xi}_1(t), \ldots, \boldsymbol{\xi}_n(t)$ are linearly independent.

Now let $\boldsymbol{y}(t)$ be a solution of $\dot{\boldsymbol{x}} = A\boldsymbol{x}$ which, when evaluated at $t = 0$, gives $\boldsymbol{y}(0) = \boldsymbol{b}$. Since the vectors $\boldsymbol{b}_i$ are a basis for $\mathscr{C}_n$ (or $\mathscr{R}_n$), there exist constants $\beta_1, \beta_2, \ldots, \beta_n$ such that

$$\boldsymbol{y}(0) = \sum_{i=1}^{n} \beta_i \boldsymbol{b}_i .$$

Now consider the vector function

$$\boldsymbol{z}(t) = \sum_{i=1}^{n} \beta_i \xi_i(t).$$

Clearly, $\boldsymbol{z}(0) = \boldsymbol{b} = \boldsymbol{y}(0)$, and $\boldsymbol{z}(t)$ is a solution of $\dot{\boldsymbol{x}} = A\boldsymbol{x}$ by the first part of the proof. Theorem 6.1 then implies that $\boldsymbol{z}(t) \equiv \boldsymbol{y}(t)$, and hence $\boldsymbol{y}(t)$ is a linear combination of the functions $\xi_i(t)$. The latter functions are therefore a basis for the solution space of $\dot{\boldsymbol{x}} = A\boldsymbol{x}$, and the proof is complete.

We define any set of n linearly independent solution vectors of $\dot{\boldsymbol{x}} = A\boldsymbol{x}$ to be a *fundamental set* of the differential equation. Thus, a fundamental set is a basis for the space of solution vectors.

If the matrix A of Theorem 6.2 is of simple structure, then the solutions of the differential equation are all of a particularly simple form. In this case we obtain a fundamental set by first constructing a set of n linearly independent right eigenvectors of A (cf. § 1.8), say $\boldsymbol{x}_1, \boldsymbol{x}_2, \ldots, \boldsymbol{x}_n$, and then observing that, if $\mu_1, \mu_2, \ldots, \mu_n$ are associated eigenvalues, the vector functions

$$\xi_i(t) = \boldsymbol{x}_i e^{\mu_i t}, \qquad i = 1, 2, \ldots, n$$

are linearly independent ($e^{\mu_i t} \neq 0$ for all finite t) and are therefore a fundamental set.

Note that A may have multiple eigenvalues and yet be of simple structure. Construction of a fundamental set when A is not simple is rather more involved, and the reader is referred to Coddington and Levinson [1] for the general analysis.

Returning to the case where A is simple, we observe that any solution of the differential equation is a linear combination of the fundamental set and can therefore be written

$$\boldsymbol{x}(t) = \sum_{i=1}^{n} c_i \boldsymbol{x}_i e^{\mu_i t}$$

for some set of constants $c_1, c_2, \ldots, c_n$. If we write $\mathbf{c}' = [c_1, c_2, \ldots, c_n]$, $X = [\boldsymbol{x}_1, \boldsymbol{x}_2, \ldots, \boldsymbol{x}_n]$ and $U = \operatorname{diag}\{\mu_1, \mu_2, \ldots, \mu_n\}$, then this *general solution* of the homogeneous equation can be written

$$\boldsymbol{x}(t) = Xe^{Ut}\,\boldsymbol{c}. \tag{6.2.1}$$

The results just obtained will now generalize immediately to the system of n differential equations of the form

$$A\dot{\boldsymbol{x}} + C\boldsymbol{x} = \boldsymbol{0}, \tag{6.2.2}$$

where A, C are $n \times n$ matrices and A is non-singular. The solutions of (6.2.2) obviously coincide with those of

$$\dot{\boldsymbol{x}} = -(A^{-1}C)\,\boldsymbol{x},$$

and we deduce from Theorem 6.2:

LEMMA *If $A\lambda + C$ is a regular pencil of matrices of order n, then the set of all solutions of* (6.2.2) *is an n-dimensional vector space.*

We now consider the n simultaneous equations represented by

$$D_l(\mathscr{D})\,\boldsymbol{q} = (A_0\mathscr{D}^l + A_1\mathscr{D}^{l-1} + \cdots + A_{l-1}\mathscr{D} + A_l)\,\boldsymbol{q} = \boldsymbol{0}, \tag{6.2.3}$$

where $A_0, A_1, \ldots, A_l$ are $n \times n$ matrices of constants and $\boldsymbol{q}(t)$ is an nth order vector function, and prove:

THEOREM 6.3 *If $D_l(\lambda)$ is a regular λ-matrix, then the solutions of* (6.2.3) *form an n-dimensional vector space. If, in addition, $D_l(\lambda)$ is a simple λ-matrix, then the solution space of* (6.2.3) *is spanned by all vectors of the type $\boldsymbol{q}e^{\lambda t}$ where $\boldsymbol{q}$ is a right latent vector of $D_l(\lambda)$ with associated latent root λ.*

Proof Let $\mathscr{A}\lambda + \mathscr{C}$ be the regular pencil of matrices of order ln associated with $\mathscr{D}_l(\lambda)$, as defined in § 4.2, and consider the first order differential equations

$$(\mathscr{A}\mathscr{D} + \mathscr{C})\,\boldsymbol{q}^{+} = \boldsymbol{0}. \tag{6.2.4}$$

Let $\boldsymbol{q}^{+}$ be divided into l partitions of order n as follows:

$$\boldsymbol{q}^{+}(t) = \begin{bmatrix} \boldsymbol{q}^{(l-1)}(t) \\ \vdots \\ \boldsymbol{q}^{(1)}(t) \\ \boldsymbol{q}^{(0)}(t) \end{bmatrix}. \tag{6.2.5}$$

Then, noting the definitions of $\mathscr{A}$ and $\mathscr{C}$, it is easily verified that

$$\boldsymbol{q}^{(s)} = \mathscr{D}^s\boldsymbol{q}^{(0)}, \qquad s = 1, 2, \ldots, l-1,$$

and $\boldsymbol{q}^{(0)}$ is a solution of (6.2.3). Conservely, if $\boldsymbol{q}^{(0)}$ is a solution of (6.2.3), then the related $\boldsymbol{q}^{+}$ is a solution of (6.2.4). Thus, we have a one-to-one correspondence between solutions of (6.2.3) and (6.2.4) The lemma implies that the solutions of (6.2.4) form a vector space of dimension ln. Therefore there exists a fundamental set

$$\boldsymbol{q}_1^{+}(t),\ \boldsymbol{q}_2^{+}(t),\ \ldots,\ \boldsymbol{q}_{ln}^{+}(t)$$

for (6.2.4) and the matrix $[\boldsymbol{q}_1^{+}, \boldsymbol{q}_2^{+}, \ldots, \boldsymbol{q}_{ln}^{+}]$ has rank ln for all finite t. Observing the partitions of (6.2.5), we now deduce that the bottom n rows of this matrix must be linearly independent. That is, the matrix

$$[\boldsymbol{q}_1^{(0)}, \boldsymbol{q}_2^{(0)}, \ldots, \boldsymbol{q}_{ln}^{(0)}]$$

of solutions of (6.2.3) has rank n for all finite t, and there exist n linearly independent solutions for (6.2.3). The solutions of (6.2.3) are easily seen to constitute a vector space, and the first statement is proved.

If $\mathscr{D}_l(\lambda)$ is simple, then so is $\mathscr{A}\lambda + \mathscr{C}$, and it is easily verified that the vectors

$$\boldsymbol{q}_i^{+}(t) = \begin{bmatrix} \lambda_i^{l-1}\boldsymbol{q}_i \\ \vdots \\ \lambda_i\boldsymbol{q}_i \\ \boldsymbol{q}_i \end{bmatrix} e^{\lambda_i t}, \qquad i = 1, 2, \ldots, ln,$$

are a fundamental set for the pencil, where the $\boldsymbol{q}_i$ are the columns of the matrix Q of latent vectors as defined in (4.3.1), and the λ_i are the associated latent roots. In the first part of the proof, the vectors $\boldsymbol{q}_i^{(0)}(t)$ which span the solution space are obviously given by

$$\boldsymbol{q}_i^{(0)}(t) = \boldsymbol{q}_i e^{\lambda_i t}, \qquad i = 1, 2, \ldots, ln,$$

and the theorem is proved.

This theorem implies that, for any solution vector $\boldsymbol{q}(t)$ of (6.2.3), there exist constants $c_1, c_2, \ldots, c_{ln}$ (not necessarily unique) such that

$$\boldsymbol{q}(t) = \sum_{i=1}^{ln} c_i \boldsymbol{q}_i e^{\lambda_i t}.$$

Defining $\mathbf{c}' = [c_1, c_2, \ldots, c_{ln}]$, this expression is abbreviated and the general solution written:

$$\boldsymbol{q}(t) = Qe^{\Lambda t}\mathbf{c}. \qquad (6.2.6)$$

Note that, since $\Lambda = \operatorname{diag}\{\lambda_1, \lambda_2, \ldots, \lambda_{ln}\}$, the definition (1.3.12) implies that $e^{\Lambda t}$ is the matrix

$$\operatorname{diag}\{e^{\lambda_1 t}, e^{\lambda_2 t}, \ldots, e^{\lambda_{ln} t}\}.$$

We now turn our attention to the non-homogeneous equation

$$D_l(\mathscr{D})\, \boldsymbol{q}(t) = \boldsymbol{f}(t), \tag{6.2.7}$$

where $\boldsymbol{f}(t)$ is a given function which we assume to be sufficiently well behaved to ensure the existence of subsequent integrals. The following familiar theorem is most important—the proof may be supplied by the reader:

THEOREM 6.4 (a) *If* $\boldsymbol{q}_1, \boldsymbol{q}_2$ *are any two distinct solutions of* (6.2.7), *then* $(\boldsymbol{q}_1 - \boldsymbol{q}_2)$ *is a solution of* (6.2.3). (b) *If* $\boldsymbol{q}_1$ *is a solution of* (6.2.7), *then so is* $\boldsymbol{q}_1 + \boldsymbol{q}$, *where* $\boldsymbol{q}$ *is an arbitrary solution of* (6.2.3).

The results of Chapter 4 allow us to obtain a solution of (6.2.7) very quickly; it may be written in the symbolic form

$$\boldsymbol{q}(t) = [D_l(\mathscr{D})]^{-1} \boldsymbol{f}(t).$$

If $\mathscr{D}_l(\lambda)$ is a simple λ-matrix, then the latent vectors may be defined as described in Theorem 4.5 and, using Theorem 4.3, we write

$$\boldsymbol{q}(t) = Q(I\mathscr{D} - \Lambda)^{-1} R' \boldsymbol{f}(t), \tag{6.2.8}$$

and using the integral (6.1.2) with $t_0 = 0$,

$$\boldsymbol{q}(t) = Q e^{\Lambda t} \int_0^t e^{-\Lambda \tau} R' \boldsymbol{f}(\tau)\, d\tau. \tag{6.2.9}$$

In order to justify the process the reader has only to verify that this expression is indeed a solution of (6.2.7).

Using Theorem 6.4 and the general solution (6.2.6), we may now say that, if $\mathscr{D}_l(\lambda)$ is a simple λ-matrix, then every solution of (6.2.7) can be written in the form

$$\boldsymbol{q}(t) = Q e^{\Lambda t} \left\{\boldsymbol{c} + \int_0^t e^{-\Lambda t} R' \boldsymbol{f}(\tau)\, d\tau\right\}. \tag{6.2.10}$$

The vector $\boldsymbol{c}$ is obviously determined so that $\boldsymbol{q}(0) = Q\boldsymbol{c}$. If any of the first $l - 1$ derivatives of the solution are required we may use equation (4.3.3) to obtain

$$\mathscr{D}^r \boldsymbol{q}(t) = Q \Lambda^r e^{\Lambda t} \left\{\boldsymbol{c} + \int_0^t e^{-\Lambda \tau} R' \boldsymbol{f}(\tau)\, d\tau\right\} \tag{6.2.11}$$

for $r = 0, 1, 2, \ldots, l-1$. Alternatively, this result may be obtained by differentiating (6.2.10) repeatedly and using (4.4.9).

If we employ (4.3.14) rather than (4.3.3), then (6.2.11) becomes

$$\mathscr{D}^r \boldsymbol{q}(t) = \sum_{i=1}^{ln} \lambda_i^r e^{\lambda_i t} \{c_i \boldsymbol{q}_i + F_i \int_0^t e^{-\lambda_i \tau} \boldsymbol{f}(\tau)\, d\tau\},$$

or, recalling the definition (4.3.13),

$$\mathscr{D}^r \boldsymbol{q}(t) = \sum_{i=1}^{ln} \lambda_i^r e^{\lambda_i t} \boldsymbol{q}_i \{c_i + \int_0^t e^{-\lambda_i \tau} \boldsymbol{r}_i' \boldsymbol{f}(\tau)\, d\tau\}. \tag{6.2.12}$$

This result can be expressed more concisely by defining the vector $\boldsymbol{\varphi}(t)$ with elements

$$\varphi_i(t) = \int_0^t e^{-\lambda_i \tau} \boldsymbol{r}_i' \boldsymbol{f}(\tau)\, d\tau, \qquad i = 1, 2, \ldots, ln, \tag{6.2.13}$$

in which case

$$\mathscr{D}^r \boldsymbol{q}(t) = Q\Lambda^r e^{\Lambda t} \{\boldsymbol{c} + \boldsymbol{\varphi}(t)\} \tag{6.2.14}$$

for $r = 0, 1, 2, \ldots, l-1$. It is apparent that, with these definitions, $\boldsymbol{\varphi}(0) = \boldsymbol{0}$ and

$$[\mathscr{D}^r \boldsymbol{q}]_{t=0} = Q\Lambda^r \boldsymbol{c}. \tag{6.2.15}$$

The contribution $Qe^{\Lambda t}\boldsymbol{c}$ to the solution (6.2.10) is often known as the *complementary function* of the differential equation (6.2.7), and the remaining part of the solution is known as a *particular integral.*

6.3 The Particular Integral when $\boldsymbol{f}(t)$ is Exponential

In many applications the vector $\boldsymbol{f}(t)$ of eqns. (6.2.7) can be written in the form

$$\boldsymbol{f}(t) = \boldsymbol{f}_0 e^{\theta t},$$

or a sum of such terms, where $\boldsymbol{f}_0$ and θ are real or complex constants. In this case (6.2.8) gives the particular solution

$$\boldsymbol{q} = Q(I\mathscr{D} - \Lambda)^{-1} R' \boldsymbol{f}_0 e^{\theta t}.$$

If θ is not equal to a latent root of $D_l(\lambda)$ and we use the indefinite integral (6.1.3), then we obtain

$$\boldsymbol{q} = e^{\theta t} Q(I\theta - \Lambda)^{-1} R' \boldsymbol{f}_0. \tag{6.3.1}$$

Or in summation form,

$$\boldsymbol{q} = e^{\theta t} \sum_{i=1}^{ln} \frac{F_i}{\theta - \lambda_i} \boldsymbol{f}_0,$$

$$= e^{\theta t} \sum_{i=1}^{ln} \frac{(\boldsymbol{r}_i' \boldsymbol{f}_0)}{\theta - \lambda_i} \boldsymbol{q}_i. \tag{6.3.2}$$

If $\theta = \lambda_m$, one of the latent roots, and H_m is the sum of the matrices F_i associated with λ_m, then (by the same means) we obtain the solution

$$\boldsymbol{q} = e^{\theta t} \sum{}' \frac{(\boldsymbol{r}_i' \boldsymbol{f}_0)}{\theta - \lambda_i} \boldsymbol{q}_i + t e^{\theta t} H_m \boldsymbol{f}_0, \tag{6.3.3}$$

where $\sum'$ denotes a summation from $i = 1$ to ln excluding those values for which $\lambda_i = \lambda_m$. If λ_m is a simple latent root, then the last term of (6.3.3) is simply

$$t e^{\theta t} (\boldsymbol{r}_m' \boldsymbol{f}_0) \boldsymbol{q}_m. \tag{6.3.4}$$

6.4 One-point Boundary Conditions

The general solution (6.2.14) of eqns. (6.2.7) contains ln undefined constants, and we now assume that there are given a sufficient number of conditions on the solution to determine these constants. We also assume that these conditions are given in the form of *ln linear* algebraic equations in the values of the variables q_i and their derivatives with respect to t evaluated at a given value of t. We lose no generality if we take this value to be $t = 0$. What we have described are linear *one*-point boundary conditions. If the relations between the q_i and their derivatives were divided among m distinct values of t then we would have m-point boundary conditions.

Linear one-point boundary conditions arise very frequently in practice and allow of an easily formulated solution (as we shall see), in contrast to the other problems for which general analysis is impractical, and among which each problem must be treated on its own merits.

We will use the obvious notation:

$$\mathscr{D}_0^r \boldsymbol{q} = \left\{ \frac{\partial^r \boldsymbol{q}}{\partial t^r} \right\}_{t=0},$$

for integer values of r, and now restrict our attention to problems having linear one-point boundary conditions. We may assume that the derivatives involved in these conditions are of order $l - 1$ at most, for otherwise, we could use equation (6.2.7) repeatedly to reduce the order to $l - 1$. With this assumption the boundary conditions may be written in the matrix form:

$$B_1\mathscr{D}_0^{l-1}\boldsymbol{q} + \cdots + B_{l-1}\mathscr{D}_0\boldsymbol{q} + B_l\boldsymbol{q}(0) = \boldsymbol{b}, \qquad (6.4.1)$$

where $B_1, B_2, \ldots, B_l$ are given $ln \times n$ matrices of constants and $\boldsymbol{b}$ is a given constant vector of order ln. We now define the $ln \times n$ λ-matrix

$$B(\lambda) = B_1\lambda^{l-1} + B_2\lambda^{l-2} + \cdots + B_{l-1}\lambda + B_l. \qquad (6.4.2)$$

We employ the general solution in the form (6.2.14):

$$\boldsymbol{q}(t) = Qe^{\Lambda t}\{\boldsymbol{c} + \boldsymbol{\varphi}(t)\}, \qquad (6.4.3)$$

and our problem is now to express $\boldsymbol{c}$, and hence $\boldsymbol{q}$, in terms of the given $B_1, B_2, \ldots, B_l$ and $\boldsymbol{b}$. If we operate on the general solution with $B(\mathscr{D})$ we obtain

$$B(\mathscr{D})\,\boldsymbol{q} = B_1\mathscr{D}^{l-1}\boldsymbol{q} + \cdots + B_{l-1}\mathscr{D}\boldsymbol{q} + B_l\boldsymbol{q},$$

and using (6.2.14) for $r = 0, 1, 2, \ldots, l-1$, we get (cf. (4.3.17)):

$$B(\mathscr{D})\boldsymbol{q} = (B_1Q\Lambda^{l-1} + \cdots + B_{l-1}Q\Lambda + B_lQ)e^{\Lambda t}\{\boldsymbol{c} + \boldsymbol{\varphi}(t)\},$$

Now at $t = 0$, $\boldsymbol{\varphi} = \boldsymbol{0}$, and, from (6.4.1), $B(\mathscr{D}_0)\,\boldsymbol{q} = \boldsymbol{b}$, so that

$$(B_1Q\Lambda_l^{-1} + \cdots + B_{l-1}Q\Lambda + B_lQ)\,\boldsymbol{c} = \boldsymbol{b} \qquad (6.4.4)$$

Denote the $ln \times ln$ matrix on the left by K. Then the existence and uniqueness of solutions of

$$D_l(\mathscr{D})\,\boldsymbol{q}(t) = \boldsymbol{f}(t) \qquad \text{with} \qquad B(\mathscr{D}_0)\,\boldsymbol{q} = \boldsymbol{b} \qquad (6.4.5)$$

depends solely on the existence and uniqueness of solution vectors $\boldsymbol{c}$ of $K\boldsymbol{c} = \boldsymbol{b}$. In particular, we deduce (in our usual notations):

THEOREM 6.5 *If $D_l(\lambda)$ is a simple λ-matrix, then* (a) *there exist solutions of the one-point boundary value problem* (6.4.5) *if and only if the ranks of the matrix K and the augmented matrix $[K\ \boldsymbol{b}]$ are the same, where*

$$K = B_1Q\Lambda^{l-1} + \cdots + B_{l-1}Q\Lambda + B_lQ, \qquad (6.4.6)$$

and (b) *the solution vector of* (6.4.5) *is unique if and only if K is non-singular.*

In case (b) of the theorem the explicit solution of the problem is

$$\boldsymbol{q}(t) = Qe^{\Lambda t}\{K^{-1}\boldsymbol{b} + \boldsymbol{\varphi}(t)\}.$$

Let us investigate a little further the conditions under which K^{-1} exists. Write $\varrho(K)$ for the rank of K, etc. By Theorem 4.4 $\varrho(Q) = n$, and the matrices B_r of (6.4.6) are $ln \times n$ so that $\varrho(B_r) \leqq n$. Thus we have $\varrho(B_rQ) \leqq n$ and $\varrho(B_rQ\Lambda^r) \leqq n$, since the matrices Λ^r are diagonal. Now in case (b) we have

$$\varrho(K) = ln = \varrho\left(\sum_{r=1}^{l} B_r Q\Lambda^{l-r}\right) \leqq \sum_{r=1}^{l} \varrho(B_rQ\Lambda^{l-r})$$

with equality only if $\varrho(B_rQ\Lambda^{l-r}) = n$ for $r = 1, 2, \ldots, l$. This would imply that $\varrho(B_rQ) = n$ and hence that $\varrho(B_r) = n$. Thus, we deduce that, *if the solution of* (6.4.5) *is unique, then the matrices* B_r, $r = 1, 2, \ldots, l$, *must each have rank* n.

If the boundary conditions take the specially simple form

$$[\mathscr{D}_0^{l-1}\boldsymbol{q}' \cdots \mathscr{D}_0\boldsymbol{q}'\ \boldsymbol{q}'(0)] = \boldsymbol{b}',$$

i.e., in which $\boldsymbol{q}$ and its first $l - 1$ derivatives are specified at $t = 0$, then it is easily shown that the matrix K reduces to the matrix $\mathscr{Q}$ of (4.3.5). In this case the solution can be written concisely in the form

$$\boldsymbol{q}(t) = Qe^{\Lambda t}\{\mathscr{Q}^{-1}\boldsymbol{b} + \boldsymbol{\varphi}(t)\}. \tag{6.4.7}$$

Such boundary conditions are described by Frazer *et al.* [1] as being of "standard type."

In practice, the inversion involved in (6.4.7) should perhaps be avoided. This can be done if we recall the first of equations (4.3.7) and write

$$\mathscr{Q}^{-1} = \mathscr{R}'\mathscr{A}.$$

Substituting for $\mathscr{A}$ and $\mathscr{R}$ from (4.2.7) and (4.3.5) and writing $\mathscr{D}_0^r\boldsymbol{q} = \boldsymbol{b}_r$, $r = 0, 1, \ldots, l-1$, yields:

$$\left.\begin{aligned}\mathscr{Q}^{-1}\boldsymbol{b} = R'A_0\boldsymbol{b}_{l-1} &+ (\Lambda R'A_0 + R'A_1)\,\boldsymbol{b}_{l-2}\\ &+ (\Lambda^2R'A_0 + \Lambda R'A_1 + R'A_2)\,\boldsymbol{b}_{l-3}\\ &\quad\vdots\\ &+ (\Lambda^{l-1}R'A_0 + \cdots + \Lambda R'A_{l-2} + R'A_{l-1})\,\boldsymbol{b}_0\end{aligned}\right\}. \tag{6.4.8}$$

6.5 The Laplace Transform Method

Historically, development of the Laplace transform technique was stimulated by problems of the kind we are presently considering. The history of the subject is fascinating and is outlined by Carslaw and Jaeger [1]. We continue to study the problem (6.4.5) though the nature of the method restricts our analysis to problems of "standard type." When $D_l(\lambda)$ is simple we already have the solution (6.4.7) for such problems, and we continue to make this assumption. Carslaw and Jaeger treat the problem in which $D_l(\lambda)$ is regular but not necessarily simple.

We will see that no new results are obtained by the use of this method. Although the technique is of very great interest in itself and is indispensable in many other problems, the motivation for our present exposition is first that it seems to be historically apt, and second habitual transformers may need to be convinced that nothing is gained in comparison with our earlier solution. We first make a formal statement of an inversion theorem for the Laplace transform.

Let $x(t)$ be bounded, piecewise continuous, and have a finite number of maxima and minima in any finite interval of the positive real axis. Suppose also that $|x(t)| = 0(e^{ct})$ as $t \to \infty$, then if

$$\bar{x}(z) = \int_0^\infty e^{-zt}\, x(t)\, dt, \qquad \mathscr{R}(z) \geqq \gamma > \max\,(c, 0), \tag{6.5.1}$$

then

$$x(t) = \frac{1}{2\pi i}\int_{\gamma - i\infty}^{\gamma + i\infty} e^{zt}\bar{x}(z)\, dz. \tag{6.5.2}$$

Equation (6.5.1) defines the Laplace transform $\bar{x}$ of x, and (6.5.2) indicates how x may be retrieved from knowledge of its transform. We observe that, if $x(t)$ also has an rth derivate $x^{(r)}(t)$, then repeated integration by parts gives for the transform of the derivative:

$$\int_0^\infty e^{-zt}x^{(r)}(t)\, dt = -\,\{x^{(r-1)}(0) + zx^{(r-2)}(0) + \cdots + z^{r-1}x(0)\} + z^r\, \bar{x}(z) \tag{6.5.3}$$

We suppose now that each element of a solution vector $\boldsymbol{q}(t)$ of (6.4.5) satisfies the conditions of the theorem. Indeed, the whole process need only be formal, as the resulting solution is easily verified. We first multiply the equation

$$(A_0\mathscr{D}^l + \cdots + A_{l-1}\mathscr{D} + A_l)\, \boldsymbol{q}(t) = \boldsymbol{f}(t)$$

by e^{-zt} and integrate from zero to infinity. If (6.5.3) is applied to the derivatives of $\boldsymbol{q}$, the terms are rearranged and we write for the initial values,

$$\boldsymbol{q}^{(r)}(0) = \boldsymbol{b}_r, \qquad r = 0, 1, 2, \ldots, l-1,$$

then we obtain

$$D_l(z)\,\bar{\boldsymbol{q}}(z) = \int_0^\infty e^{-zt}\,\boldsymbol{f}(t)\,dt + \boldsymbol{a}(z), \tag{6.5.4}$$

where

$$\boldsymbol{a}(z) = A_0\boldsymbol{b}_{l-1} + (A_0 z + A_1)\,\boldsymbol{b}_{l-2} + \cdots + (A_0 z^{l-1} + \cdots + A_{l-2} z + A_{l-1})\,\boldsymbol{b}_0 \tag{6.5.5}$$

We now premultiply (6.5.4) by $[D_l(z)]^{-1}$, and on the hypothesis that $D_l(\lambda)$ is simple we use (4.3.14) with $r = 0$ and write

$$[D_l(z)]^{-1} = \sum_{k=1}^{ln} \frac{F_k}{z - \lambda_k}.$$

Thus, (6.5.4) gives

$$\bar{\boldsymbol{q}}(z) = \sum_{k=1}^{ln} \frac{F_k}{z - \lambda_k} \int_0^\infty e^{-z\tau}\,\boldsymbol{f}(\tau)\,d\tau + \sum_{k=1}^{ln} \frac{F_k}{z - \lambda_k}\,\boldsymbol{a}(z).$$

In order to apply the inversion formula (6.5.2) we multiply by e^{zt} and integrate; then

$$2\pi i\boldsymbol{q}(t) = \sum_{k=1}^{ln} F_k \int_0^\infty \boldsymbol{f}(\tau)\,d\tau \int_{\gamma - i\infty}^{\gamma + i\infty} \frac{e^{z(t-\tau)}}{z - \lambda_k}\,dz + \sum_{k=1}^{ln} F_k \int_{\gamma - i\infty}^{\gamma + i\infty} \frac{e^{zt}\boldsymbol{a}(z)}{z - \lambda_k}\,dz,$$

where the order of integration in the first term has been formally inverted. Now each of the z integrals can be evaluated by contour integration. We consider the contour $\mathscr{C}$ in the z-plane sketched in Fig. 6.1. For each of the integrands in the above equation it can be shown that the integral over the curved part of the contour tends to zero as $R \to \infty$. Thus, as $R \to \infty$ the contour integral approaches the required line integral. The contour integrals can now be evaluated in terms of the residues of the integrands and if R and γ are such that all the poles of the integrands (i.e., the points $z = \lambda_k$) are inside the contour, then the value of the integrals is invariant with R. Thus, provided $\gamma > \max \mathscr{R}(\lambda_k)$ and $R > \max |\lambda_k|$, we replace

$$\int_{\gamma - i\infty}^{\gamma + i\infty} \quad \text{by} \quad \int_{\mathscr{C}}.$$

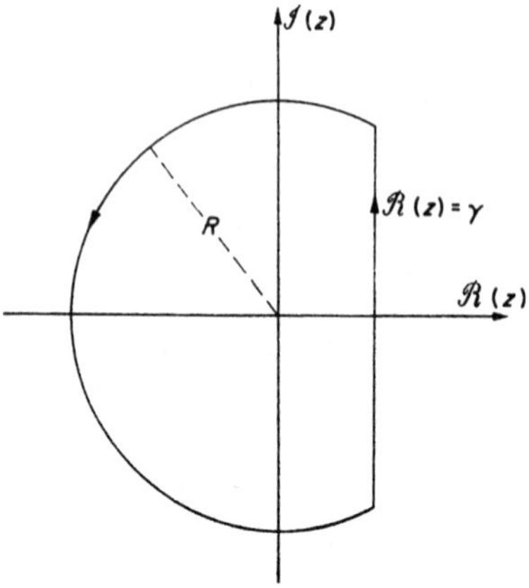

FIG. 6.1. The contour $\mathscr{C}$.

In the first and second terms of the above equations there are residues $e^{\lambda_k(t-\tau)}$ and $e^{\lambda_k(t)}\boldsymbol{a}(\lambda_k)$ respectively at $z = \lambda_k$. Thus we obtain

$$\boldsymbol{q}(t) = \sum_{k=1}^{ln} F_k e^{\lambda_k t} \int_0^\infty e^{-\lambda_k \tau} \boldsymbol{f}(\tau)\, d\tau + \sum_{k=1}^{ln} F_k e^{\lambda_k t} \boldsymbol{a}(\lambda_k).$$

Recalling that $F_i = \boldsymbol{q}_i \boldsymbol{r}_i'$ and the definition (6.5.5) of $\boldsymbol{a}(z)$, the above solution can (after some manipulation) be identified with (6.4.7) with $\mathscr{Q}^{-1}\boldsymbol{b}$ written in the form (6.4.8).

6.6 Second Order Differential Equations

In the case $l = 2$ we may, of course, use the results already obtained provided $\mathscr{D}_2(\lambda)$ is a simple λ-matrix, but, if the stronger conditions of Theorem 3.5 apply, then we may formulate a rather more flexible result; more flexible in the sense that the biorthogonalization process of Theorem 4.5 need not be carried out, and that only $2n$, instead of the full $4n$, latent vectors are required.

In this case the symbolic solution

$$\boldsymbol{q}(t) = [D_2(\mathscr{D})]^{-1} \boldsymbol{f}(t)$$

of (6.2.7) can be written in the form

$$\boldsymbol{q}(t) = Q_1(I\mathscr{D} - \Lambda_1)^{-1} T(I\mathscr{D} - \Lambda_2)^{-1} R_2 \boldsymbol{f}(t),$$

where we have used (4.6.1) and the definition (4.6.3). If we employ the technique described at the end of § 6.1 we can interpret this as

$$\boldsymbol{q}(t) = Q_1 e^{\Lambda_1 t} \int_0^t e^{-\Lambda_1 \tau} T e^{\Lambda_2 \tau} \int_0^\tau e^{-\Lambda_2 \nu} R_2' \boldsymbol{f}(\nu)\, d\nu\, d\tau .$$

Reversing the order of integration by Dirichlet's rule we obtain

$$\boldsymbol{q}(t) = Q_1 e^{\Lambda_1 t} \int_0^t K(\nu)\, e^{-\Lambda_2 \nu} R_2' \boldsymbol{f}(\nu)\, d\nu, \tag{6.6.1}$$

where

$$K(\nu) = \int_0^t e^{-\Lambda_1 \tau}\, T e^{\Lambda_2 \tau}\, d\tau .$$

The i, j element of the final integrand is $t_{ij} \exp\,(\lambda_{n+j} - \lambda_i)\,\tau$, so that the corresponding element of $K(\nu)$ is

$$\frac{1}{\lambda_{n+j} - \lambda_i} \{e^{-\lambda_i t}\, t_{ij}\, e^{\lambda_{n+j} t} - e^{-\lambda_i \nu}\, t_{ij}\, e^{\lambda_{n+j} \nu}\},$$

and if we use the definition (4.6.11) we may write

$$K(\nu) = e^{-\Lambda_1 t}\, \Omega e^{\Lambda_2 t} - e^{-\Lambda_1 \nu}\, \Omega e^{\Lambda_2 \nu}.$$

The particular integral (6.6.1) is therefore

$$\boldsymbol{q}(t) = Q_1 \int_0^t \{\Omega e^{\Lambda_2 (t-\nu)} - e^{\Lambda_1 (t-\nu)} \Omega\}\, R_2' \boldsymbol{f}(\nu)\, d\nu . \tag{6.6.2}$$

This expresses $\boldsymbol{q}$ in terms of the unnormalized column vectors of Q_1 and R_2. However, if we use the results of Theorem 4.10 and write $Q_2 = Q_1 \Omega$ and $R_1' = -\Omega R_2'$ we find that

$$\boldsymbol{q}(t) = \int_0^t \{Q_2 e^{\Lambda_2 (t-\nu)} R_2' + Q_1 e^{\Lambda_1 (t-\nu)} R_1'\}\, \boldsymbol{f}(\nu)\, d\nu, \tag{6.6.3}$$

$$= Q e^{\Lambda t} \int_0^t e^{-\Lambda \nu} R' \boldsymbol{f}(\nu)\, d\nu,$$

which is the particular integral of (6.2.9) once more. Further, if we differentiate with respect to t and then use (4.6.14) we find that

$$\mathcal{D}\boldsymbol{q} = Q \Lambda e^{\Lambda t} \int_0^t e^{-\Lambda \nu} R' \boldsymbol{f}(\nu)\, d\nu .$$

CHAPTER 7

THE THEORY OF VIBRATING SYSTEMS

7.1 Introduction

In this chapter we apply some of the results already formulated to well-known problems in the theory of vibration. We first of all formulate the energies and dissipation functions required to apply the Lagrange method for obtaining equations of motion. This is done in the following section; we formulate the problems for viscous damping forces and also for hysteretic damping forces—so far as they are defined.

We then consider the solutions for various problems of gradually increasing difficulty. Thus in § 7.3 we solve the classical problem for the undamped oscillations of a conservative system about a position of equilibrium. We also consider forced oscillations, then introduce dissipative internal forces and finally, in § 7.7, we discuss the more general problem of self-excited oscillations with some external energy source.

In all the work of this and later chapters we will employ a complex notation for the displacements of a system and for their time derivatives. A point of the system whose displacement at time t is given by

$$x \cos \omega t = \mathscr{R}(xe^{i\omega t}),$$

where x is real, has a velocity

$$\begin{aligned} -\omega x \sin \omega t &= \omega x \cos (\omega t + \tfrac{1}{2}\pi) = \mathscr{R}(\omega x e^{i\omega t}\, e^{i(1/2)\pi}) \\ &= \mathscr{R}(i\omega x e^{i\omega t}). \end{aligned}$$

Thus, if we write $p = xe^{i\omega t}$ for the displacement, but understand by this $\mathscr{R}(p)$ for the *physical* displacement, then we have for the velocity, $\dot{p} = i\omega p$, but again it is only $\mathscr{R}(\dot{p})$ which has any physical meaning. More generally, suppose that, for any real valued function φ of the time t, the displacement at a point is of the form

$$x\varphi(t)\, e^{\mu t} \cos(\omega t - \theta) = \mathscr{R}\{\xi\varphi(t)\, e^{\lambda t}\}$$

where $\xi = xe^{-i\theta}$, and $\lambda = \mu + i\omega$. Then write

$$p = \xi\varphi(t)\, e^{\lambda t}. \tag{7.1.1}$$

It can be verified that we now have $\mathscr{R}(\dot{p})$ for the velocity of the point in question.

Now consider a *real* differential operator of the form

$$F(\mathscr{D}) = a_0\mathscr{D}^l + a_1\mathscr{D}^{l-1} + \dots + a_l,$$

where $\mathscr{D} = d/dt$ and the coefficients a_r may be real valued functions of t. It is now clear that we may write

$$F(\mathscr{D})\, \mathscr{R}(p) = \mathscr{R}F(\mathscr{D})\, p.$$

Thus, the solution of a differential equation for the displacements:

$$F(\mathscr{D})\, \mathscr{R}(p) = \mathscr{R}(f),$$

where f may be any complex valued function of t, may be obtained as the real part of the solution of

$$F(\mathscr{D})\, p = f,$$

provided that we check *a posteriori* that all the solutions are of the form (7.1.1). This will always be the case in our applications.

7.2 Equations of Motion

We consider a mechanical system whose displacement from some equilibrium position can be specified by a *finite* number of coordinates. This assumption immediately precludes exact analyses of the elastic vibrations of continuous media, metal plates and shells for instance. However, the Rayleigh–Ritz technique (Gould [1]) provides a very powerful method for approximating continuous systems by others with only a finite number of coordinates. Our initial assumption is therefore not so restrictive as it may at first appear.

Let the chosen coordinates be $p_1, p_2, \dots, p_n$. We now assume that these coordinates can be varied arbitrarily and independently of each other without violating any constraints which may act on the system. This implies that the coordinates are *generalized coordinates* and that the system is *holonomic* (Whittaker [1]). We may therefore formulate equations of motion for the system by means of the classical Lagrangean method,† and in order to do this we first formulate the kinetic energy of the system and then the potential energy of the restoring forces, i.e., those conservative forces which tend to restore the system to its equilibrium position, once it is disturbed. Finally, we consider how the effects of other internal and external forces may be included in our formulation. We will denote time derivatives by the usual "dot notation."

With the above assumptions the kinetic energy, T, may be expressed as a homogeneous quadratic form in the generalized velocities $\dot{p}_j$, $j = 1, 2, \ldots, n$, and we write

$$T = \tfrac{1}{2} \sum_k \sum_j a_{jk} \dot{p}_j \dot{p}_k , \tag{7.2.1}$$

where the coefficients a_{jk} will, in general, be functions of the generalized coordinates. We now confine our attention to *small* displacements from the equilibrium position, in which case we expand the coefficients a_{ik} as power series in $p_1, \ldots, p_n$. This is accomplished by means of Taylor's theorem. On putting these expansions in (7.2.1) we assume that the generalized velocities are also small, and retain only the second order terms in the resulting expression. Thus, we retain only the constant term of each Taylor series and, to this order of approximation, we may treat the coefficients a_{jk} in (7.2.1) as constants.

The kinetic energy may also be written in the form:

$$T = \tfrac{1}{2} \dot{\mathbf{p}}' A \dot{\mathbf{p}} , \tag{7.2.2}$$

(cf. eqn. (1.5.3), where $\dot{\mathbf{p}}$ is the vector with the generalized velocities as components, and A is a real symmetric matrix. In addition, T is always positive, whatever the velocities may be, and it follows that A is also *positive definite* (cf. § 1.5).

If we consider an elastic system, then provided the displacements are small enough, the elastic restoring forces are linear and conservative and may be derived from a potential function. More generally, we include in the potential function $V(p_1, \ldots, p_n)$ those conservative forces which tend to restore the system to its

† Being more concerned with the matrix formalism than with basic concepts of vibration theory some readers may find our treatment rather superficial. Such readers may be referred to Bishop and Johnson [1] or Pipes [1] for a more leisurely development of the theory.

equilibrium position whenever the coordinates are slightly varied. Now V may be expanded as a Taylor series in the coordinates, giving

$$V(p_1, \ldots, p_n) = V_0 + \sum_{j=1}^{n} b_j p_j + \tfrac{1}{2} \sum_j \sum_k c_{jk} p_j p_k + \cdots,$$

where V_0 is the potential energy in the equilibrium position and

$$b_j = \frac{\partial V}{\partial p_j}, \qquad c_{jk} = \frac{\partial^2 V}{\partial p_j \, \partial p_k},$$

these derivatives being evaluated at the equilibrium position. Without loss of generality we may assume $V_0 = 0$, and since V has a stationary value at a position of equilibrium, we also have $b_j = 0$, $j = 1, 2, \ldots, n$. Thus, to the accuracy with which T was formulated, we may write

$$V = \tfrac{1}{2} \sum_j \sum_k c_{jk} p_j p_k \tag{7.2.3}$$

$$= \tfrac{1}{2} \boldsymbol{p}' C \boldsymbol{p} \tag{7.2.4}$$

in an obvious notation, and C is real and symmetric.

If the position about which the motion occurs is one of *stable* equilibrium, it can be shown that V has a strict minimum at this point (in the n-dimensional space of the coordinate vectors). That is, $V > 0$ for all vectors $\boldsymbol{p}$ other than the zero vector and lying in some neighborhood of the zero vector. This implies that C is also positive definite. However, cases in which the equilibrium is neutrally stable arise quite frequently, and in this case we have $V \geqq 0$ in some neighborhood of the origin, the equality being attained for some vectors other than the zero vector. The matrix C is then non-negative definite (cf. § 1.5). We will always make the latter more general assumption.

We will also take into account internal dissipative forces in the system. Experiments suggest that, in the case of small vibrations, there are two physically sensible assumptions which may be made in this connection (Robertson and Yorgiadis [1], Lazan [1]) according as the rate at which energy is absorbed in sinusoidal motion increases linearly with the frequency, or is independent of the frequency. The former case is reproduced if we postulate a force acting at each point of the system which is proportional to the velocity at that point, and which is opposed to the motion.

This gives rise to formulation of a generalized force vector $\boldsymbol{P}$ due to these damping effects in the form

$$\boldsymbol{P} = -B\dot{\boldsymbol{p}} \tag{7.2.5}$$

where B is a real $n \times n$ matrix. The rate at which the force vector $\boldsymbol{P}$ is doing work is then given by

$$\boldsymbol{P}'\dot{\boldsymbol{p}} = -\dot{\boldsymbol{p}}'B'\dot{\boldsymbol{p}} = -\sum\sum b_{jk}\dot{p}_j\dot{p}_k\,.$$

This quantity may also be described as the rate at which the viscous forces dissipate energy, and for physical reasons is supposed never to be positive, whatever the generalized velocities may be. The matrix B is in fact determined as the matrix of the above quadratic form in the generalized velocities, and may be assumed to be symmetric and non-negative definite. We now define the *dissipation function* (Lord Rayleigh [1]) to be

$$F = \tfrac{1}{2}\dot{\boldsymbol{p}}'B\dot{\boldsymbol{p}}, \tag{7.2.6}$$

in which case the generalized force P_j is given by

$$P_j = -\frac{\partial F}{\partial \dot{p}_j}. \tag{7.2.7}$$

In order to verify this statement we note first of all that

$$\frac{\partial \boldsymbol{p}}{\partial p_j} = \frac{\partial \dot{\boldsymbol{p}}}{\partial \dot{p}_j} = \boldsymbol{e}_j, \qquad j = 1, 2, \ldots, n,$$

the unit vector with a one in the jth position and zeros elsewhere. Differentiating (7.2.6) we now obtain

$$\begin{aligned}\frac{\partial F}{\partial \dot{p}_j} &= \tfrac{1}{2}\boldsymbol{e}_j'B\dot{\boldsymbol{p}} + \tfrac{1}{2}\dot{\boldsymbol{p}}'B\boldsymbol{e}_j\\ &= \boldsymbol{e}_jB\dot{\boldsymbol{p}}\\ &= (j\text{th element of } B\dot{\boldsymbol{p}}),\end{aligned}$$

so that, referring to (7.2.5), the statement (7.2.7) is verified.

Forces giving rise to a rate of energy dissipation which is independent of the frequency will be referred to as hysteretic damping forces. In this case the resisting forces acting in *sinusoidal* motion are easily formulated, and Bishop and Johnson [1] show that they can be derived from a dissipation function

$$S = \tfrac{1}{2}i\sum_j\sum_k g_{jk}p_jp_k, \tag{7.2.8}$$

where the coefficients g_{jk} are real constants. The generalized forces are then obtained in the form

$$P_j = -\frac{\partial S}{\partial p_j}$$

and correspond to forces which are proportional to the displacement at each point but are in phase with the velocity at the point.

It is implicit in (7.2.8) that, as discussed in the previous section, we are using a complex notation for the displacements p_j. This interpretation accounts for the i factor in (7.2.8), and accentuates the fact that this equation only has significance for pure sinusoidal motion.

In general then, for *sinusoidal* motion the Lagrangean equations of motion take the form:

$$\frac{d}{dt}\left(\frac{dT}{d\dot{p}_j}\right) + \frac{\partial F}{\partial \dot{p}_j} + \frac{\partial S}{\partial p_j} + \frac{\partial V}{\partial p_j} = P_j, \qquad (7.2.9)$$

for $j = 1, 2, \ldots, n$, and where the P_j now denote any further unspecified generalized forces. We have seen above that

$$\frac{\partial F}{\partial \dot{p}_j} = (j\text{th element of } B\dot{\boldsymbol{p}})$$

so that the vector with elements $\partial F/\partial \dot{p}_j$, $j = 1, 2, \ldots, n$, is simply $B\dot{\boldsymbol{p}}$. Using a similar analysis for the derivatives of the other quadratic forms it is found that the equations of motion may be written in the matrix form:

$$A\ddot{\boldsymbol{p}} + B\dot{\boldsymbol{p}} + (C + iG)\,\boldsymbol{p} = \boldsymbol{P}. \qquad (7.2.10)$$

If the frequency of the sinusoidal motion is ω we can write $\dot{\boldsymbol{p}} = i\omega\boldsymbol{p}$ and obtain

$$A\ddot{\boldsymbol{p}} + \left(B + \frac{1}{\omega}G\right)\dot{\boldsymbol{p}} + C\boldsymbol{p} = \boldsymbol{P}.$$

We see that the matricial differential operator involved is now real, so that the complex notation discussed in § 7.1 still has significance.

If it is not known *a priori* that the solution will represent a sinusoidal motion, then S cannot appear in (7.2.9) and solutions are then sought for the equations

$$A\ddot{\boldsymbol{p}} + B\dot{\boldsymbol{p}} + C\boldsymbol{p} = \boldsymbol{P}. \qquad (7.2.11)$$

These are the classical equations investigated at some length by Lord Rayleigh, and by many others.

7.3 Solutions under the Action of Conservative Restoring Forces Only

We consider now some special solutions for *undamped* systems, that is, systems in which the viscous and hysteretic damping forces discussed in the previous section are absent, or may be ignored.

Consider first of all problems in which the vector $\boldsymbol{P}$ vanishes identically and we are left with the homogeneous equations of free vibration:

$$A\ddot{\boldsymbol{p}} + C\boldsymbol{p} = \boldsymbol{0}. \tag{7.3.1}$$

We look for sinusoidal solutions which we represent by the real part of

$$\boldsymbol{p} = \boldsymbol{q}e^{i\omega t} \tag{7.3.2}$$

where $\boldsymbol{q}$ is independent of t. Writing $\lambda = -\omega^2$, equation (7.3.1) leads to the latent root problem determined by

$$(A\lambda + C)\,\boldsymbol{q} = \boldsymbol{0}, \tag{7.3.3}$$

where $A\lambda + C$ is a real, regular, and symmetric matrix pencil as defined in § 2.1. Furthermore, A is positive definite so that the conditions of Theorem 2.11 are satisfied and $A\lambda + C$ is therefore a simple matrix pencil. From the corollary to that theorem we see that all the latent roots and vectors are real.

In addition, we easily deduce from (7.3.3) that for any latent root λ with a latent vector $\boldsymbol{q}$:

$$\lambda = -\frac{\boldsymbol{q}'C\boldsymbol{q}}{\boldsymbol{q}'A\boldsymbol{q}},$$

and since C is non-negative definite we conclude that *all the latent roots are real and do not exceed zero.*

Now $\lambda = -\omega^2$, so that we obtain n real *natural frequencies* of oscillation from the latent roots. Since we are only interested in the real part of the solution (7.3.2) the apparent ambiguity in the sign of ω is not relevant.

Let the natural frequencies be ω_j ($j = 1, 2, \ldots, n$), not necessarily distinct, then we can construct a set of n *real*, linearly independent latent vectors $\boldsymbol{q}_j$ for (7.3.3) and the n solutions $\boldsymbol{p}_j = \boldsymbol{q}_j e^{i\omega_j t}$ form a fundamental set for the equation (7.3.1) (cf. § 6.2), and *all solutions of eqns.* (7.3.1) *may be expressed as linear combinations of these n fundamental solutions.* Notice also that the solution $\boldsymbol{p}_j(t)$ represents a simple harmonic motion in which all the particles of the system move together in the same phase.

Having observed that $A\lambda + C$ is a simple pencil of matrices we learn from Theorem 2.1 and equation (2.2.3) that the vectors $\boldsymbol{q}_j$ may be so defined that

$$\boldsymbol{q}_j' A \boldsymbol{q}_k = \delta_{jk}, \qquad \text{and} \qquad \boldsymbol{q}_j' C \boldsymbol{q}_k = \omega_j^2 \delta_{jk}. \tag{7.3.4}$$

When these conditions are satisfied we will say that the vectors $\boldsymbol{q}_j$ define *natural modes of vibration* for the undamped system. We will call the vectors $\boldsymbol{q}_j$ *principal vectors.*

We deduce from (2.8.4) that, if $\omega_1 \leqq \omega_2 \leqq \dots \leqq \omega_n$, then the Rayleigh quotient is such that

$$\omega_1^2 \leqq \frac{\boldsymbol{q}' C \boldsymbol{q}}{\boldsymbol{q}' A \boldsymbol{q}} \leqq \omega_n^2,$$

for any real vector $\boldsymbol{q} \neq \boldsymbol{0}$.

If we define the $n \times n$ non-singular matrix Q to be the matrix with columns $\boldsymbol{q}_1, \boldsymbol{q}_2, \dots, \boldsymbol{q}_n$, and $W = \text{diag}\,\{\omega_1, \omega_2, \dots, \omega_n\}$, then eqns. (7.3.4) can be written:

$$Q'AQ = I, \qquad \text{and} \qquad Q'CQ = W^2. \tag{7.3.5}$$

Let us now express eqns. (7.3.1) in terms of new coordinates $\xi_1, \xi_2, \dots, \xi_n$ defined by the transformation

$$\boldsymbol{\xi} = Q^{-1}\boldsymbol{p}, \tag{7.3.6}$$

with the inverse transformation $\boldsymbol{q} = Q\boldsymbol{\xi}$. Substituting this in (7.3.1), premultiplying the equation by Q', and then using (7.3.5) leads to the result:

$$\ddot{\boldsymbol{\xi}} + W^2 \boldsymbol{\xi} = \boldsymbol{0},$$

or,

$$\ddot{\xi}_j + \omega_j^2 \xi_j = 0, \qquad j = 1, 2, \dots, n.$$

Thus, the *principal coordinates* defined by the transformation (7.3.6) are such that motion may occur in which only one coordinate varies, ξ_j, say, and if $\omega_j \neq 0$ the motion is then simple harmonic having a natural frequency ω_j.

From (7.2.2) and (7.2.4) we also see that the kinetic and potential energies, when expressed in terms of the principal coordinates, have the form:

$$T = \tfrac{1}{2}(\dot{\xi}_1^2 + \dot{\xi}_2^2 + \cdots + \dot{\xi}_n^2)$$
$$V = \tfrac{1}{2}(\omega_1^2\xi_1^2 + \omega_2^2\xi_2^2 + \cdots + \omega_n^2\xi_n^2).$$

7.4 The Inhomogeneous Case

We consider now the problem of the previous section with the addition of prescribed generalized forces,

$$P_j = f_j(t), \qquad j = 1, 2, \ldots, n, \tag{7.4.1}$$

or $\boldsymbol{P} = \boldsymbol{f}(t)$, appearing on the right of eqns. (7.3.1). Thus,

$$A\ddot{\boldsymbol{p}} + C\boldsymbol{p} = \boldsymbol{f}(t). \tag{7.4.2}$$

If we make the transformation to principal coordinates, premultiply the resulting equation by Q', and use (7.3.5) again; we find that

$$\ddot{\boldsymbol{\xi}} + W^2\boldsymbol{\xi} = Q'\boldsymbol{f}(t), \tag{7.4.3}$$

which can be written

$$\ddot{\xi}_j + \omega_j^2\xi_j = \boldsymbol{q}_j'\boldsymbol{f}(t), \qquad j = 1, 2, \ldots, n. \tag{7.4.4}$$

We can now obtain particular integrals of these equations in the classical forms:

$$\xi_j(t) = \frac{1}{\omega_j}\int_0^t \{\boldsymbol{q}_j'\boldsymbol{f}(\nu)\} \sin\omega_j(t-\nu)\,d\nu, \qquad \text{if} \qquad \omega_j \neq 0, \tag{7.4.5}$$

and

$$\xi_j(t) = \int_0^t\int_0^\nu \{\boldsymbol{q}_j'\boldsymbol{f}(\mu)\}\,d\mu\,d\nu$$
$$= \int_0^t \{\boldsymbol{q}_j'\boldsymbol{f}(\mu)\}(t-\mu)\,d\mu, \qquad \text{if} \qquad \omega_j = 0. \tag{7.4.6}$$

If we suppose, for simplicity, that $\omega_j \neq 0$ for all n values of j, then

$$\boldsymbol{\xi}(t) = W^{-1}\int_0^t \{\sin W(t-\nu)\}\,Q'\boldsymbol{f}(\nu)\,d\nu,$$

and the solution for the generalized coordinates is obtained from (7.3.6):

$$\boldsymbol{p}(t) = QW^{-1}\int_0^t \{\sin W(t-\nu)\}\,Q'\boldsymbol{f}(\nu)\,d\nu, \tag{7.4.7}$$

or,

$$\boldsymbol{p}(t) = \sum_{j=1}^{n} \boldsymbol{q}_j \frac{1}{\omega_j} \int_0^t \{\boldsymbol{q}_j'\boldsymbol{f}(\nu)\} \sin \omega_j(t - \nu)\, d\nu, \tag{7.4.8}$$

whichever form is preferred. These results can also be obtained from (6.6.2), and it is instructive to make this application.

If, in particular, the function $\boldsymbol{f}(t)$ is given by

$$\boldsymbol{f}(t) = \boldsymbol{f}_0 e^{i\omega t},$$

where $\boldsymbol{f}_0$ and ω are real, then it is easily shown that that part of the solution (7.4.3) which corresponds to the response of the system at the *forcing frequency* ω is obtained in the form

$$\boldsymbol{p}(t) = e^{i\omega t} \sum_{j=1}^{n} \boldsymbol{q}_j \frac{\omega_j}{\omega_j^2 - \omega^2} (\boldsymbol{q}_j'\boldsymbol{f}_0). \tag{7.4.9}$$

This result illustrates the *resonance* phenomenon, for, if we imagine ω to be varied continuously in a neighborhood of a natural frequency ω_j, (7.4.9) indicates that in general $\boldsymbol{p} \to \infty$ as $\omega \to \omega_j$. However, the singularity may be avoided if $(\boldsymbol{q}_j'\boldsymbol{f}_0) = 0$ for all the principal vectors associated with ω_j, i.e., if $\boldsymbol{f}_0$ is orthogonal to the subspace spanned by the principal vectors (or latent vectors) associated with ω_j.

7.5 Solutions Including the Effects of Viscous Internal Forces

We consider first the homogeneous equations obtained from (7.2.11) by putting $\boldsymbol{P} = \boldsymbol{0}$. That is we permit *viscous* dissipative forces and assume that there are no generalized forces present. Thus, as noted in § 7.2, A, B, C are real and symmetric, A is positive definite, B and C are non-negative definite, and

$$A\ddot{\boldsymbol{p}} + B\dot{\boldsymbol{p}} + C\boldsymbol{p} = \boldsymbol{0}. \tag{7.5.1}$$

We now refer to Chapter 6 for the appropriate solutions of this and succeeding equations, and we *assume* that the matrix

$$D_2(\lambda) = A\lambda^2 + B\lambda + C \tag{7.5.2}$$

is a *simple* λ-matrix, with latent roots $\lambda_1, \lambda_2, \ldots, \lambda_{2n}$. Clearly, if any of these roots are complex, then they will arise in conjugate pairs. Since $D_2(\lambda)$ is symmetric, the right and left latent vectors as defined in (4.3.1) and (4.3.2) may be assumed to coincide and they are denoted by $\boldsymbol{q}_1, \boldsymbol{q}_2, \ldots, \boldsymbol{q}_{2n}$. These vectors are real, or arise in complex conjugate pairs, according as the associated latent roots are real or complex, and we say that they define the *free modes of vibration* for the damped system. This terminology will be justified in § 7.9. We have proved in Theorem 6.3 that the vectors

$$\boldsymbol{p}_j(t) = e^{\lambda_j t}\boldsymbol{q}_j, \qquad j = 1, 2, \ldots, 2n, \tag{7.5.3}$$

span the solution space of (7.5.1), and any solution of this equation can therefore be written in the form

$$\boldsymbol{p}(t) = Qe^{\Lambda t}\,\boldsymbol{c}, \tag{7.5.4}$$

where $\boldsymbol{c}$ is a column vector of $2n$ arbitrary constants.

We define a general solution of (7.5.1) to be stable, neutrally stable, or unstable according as the real parts of the latent roots appearing in (7.5.4) are (i) all negative, (ii) include at least one which is zero, the remainder being negative, or, (iii) include at least one which is positive.

If some of the latent roots appearing in (7.5.4) are real, then, of course, "the real part" refers to the roots themselves.

THEOREM 7.1 *The general solution of* (7.5.1) *is stable or neutrally stable, and it is stable if B and C are positive definite.*

Proof Let λ, $\boldsymbol{q}$ be a set of latent values of $D_2(\lambda)$; then

$$(A\lambda^2 + B\lambda + C)\,\boldsymbol{q} = \boldsymbol{0}. \tag{7.5.5}$$

Premultiplying by $\bar{\boldsymbol{q}}$, we observe that $a = \bar{\boldsymbol{q}}'A\boldsymbol{q}$ is real, and defining b and c similarly we have

$$a\lambda^2 + b\lambda + c = 0. \tag{7.5.6}$$

Now by equation (1.5.7) we also have $a > 0$ and $b, c \geqq 0$. For all *real* latent roots the theorem now follows by inspection of (7.5.6), for the left-hand side can only vanish (for non-zero λ) if $\lambda < 0$. If C is singular, then there exists at least one null latent root, and if there exists a null latent root then $c = 0$ for some $\boldsymbol{q}$, which implies that C is non-negative definite and must therefore be singular.

For the complex latent roots we first write the solutions of (7.5.6) in the form

$$2a\lambda = -b \pm (b^2 - 4ac)^{1/2} = -b \pm i(4ac - b^2)^{1/2} \tag{7.5.7}$$

and observe that

$$\mathscr{R}(\lambda) = -(b/2a) \leqq 0.$$

The strict inequality will hold if b is always positive and, in particular, if B is positive definite. This theorem implies that all latent roots of (7.5.2) lie in the left-hand half of the complex λ-plane. However, much stronger results can be obtained describing the location of the latent roots, and these will be pursued in Chapter 9.

Consider now the effect of introducing prescribed external forces represented by the vector $\boldsymbol{f}(t)$, as in (7.4.2). We now have

$$A\ddot{\boldsymbol{p}} + B\dot{\boldsymbol{p}} + C\boldsymbol{p} = \boldsymbol{f}(t). \qquad (7.5.8)$$

For the solutions of these equations we can improve on those obtained in Chapter 6 in matters of detail only. The new features are the symmetry of the operator and the fact that it is real. We will make a further statement of the results, taking advantage of these properties. We suppose that there exist $2s$ real latent roots of $D_2(\lambda)$ and $n - s$ conjugate pairs of complex roots, all the roots being counted according to their multiplicities. Let the real roots be numbered from 1 to $2s$, and let the roots $\lambda_{n+s+1}, \ldots, \lambda_{2n}$ be the complex conjugates of $\lambda_{2s+1}, \ldots, \lambda_{n+s}$. Define

$$Q_1 = [\boldsymbol{q}_1, \boldsymbol{q}_2, \ldots, \boldsymbol{q}_{2s}], \qquad Q_2 = [\boldsymbol{q}_{2s+1}, \ldots, \boldsymbol{q}_{n+s}], \qquad (7.5.9)$$

and

$$\Lambda_1 = \operatorname{diag}\{\lambda_1, \ldots, \lambda_{2s}\}, \qquad \Lambda_2 = \operatorname{diag}\{\lambda_{2s+1}, \ldots, \lambda_{n+s}\}. \qquad (7.5.10)$$

Then, *assuming that $A\lambda^2 + B\lambda + C$ is a simple λ-matrix, the latent vectors can be defined so that, for a set of latent values λ, $\boldsymbol{q}$,*

$$\text{(i)} \qquad \boldsymbol{q}'(2A\lambda + B)\,\boldsymbol{q} = 1,$$

and (ii) *those vectors belonging to a multiple root are orthogonal with respect to $2A\lambda + B$, and when these conditions are satisfied a particular integral of* (7.5.8) (*obtained from* (6.2.9)) *is*

$$\boldsymbol{p}(t) = Q_1 e^{\Lambda_1 t} \int_0^t e^{-\Lambda_1 \tau} Q_1' \boldsymbol{f}(\tau)\, d\tau + 2 \int_0^t \mathscr{R}\{Q_2 e^{\Lambda_2(t-\tau)} Q_2'\}\, \boldsymbol{f}(\tau) d\tau. \qquad (7.5.11)$$

For the generalized velocities we have

$$\dot{\boldsymbol{p}}(t) = Q_1 \Lambda_1 e^{\Lambda_1 t} \int_0^t e^{-\Lambda_1 \tau} Q_1' \boldsymbol{f}(\tau)\, d\tau + 2 \int_0^t \mathscr{R}\{Q_2 \Lambda_2 e^{\Lambda_2(t-\tau)} Q_2'\}\, \boldsymbol{f}(\tau)\, d\tau. \qquad (7.5.12)$$

A general solution is obtained by combining (7.5.11) with the general solution (7.5.4) of the homogeneous problem. Again, we may write (7.5.11) in the summation form:

$$\boldsymbol{p}(t) = \sum_{j=1}^{2s} \boldsymbol{q}_j e^{-\Lambda_j t} \int_0^t e^{-\Lambda_j \tau} \boldsymbol{q}_j' \boldsymbol{f}(\tau)\, d\tau + 2 \sum_{j=2s+1}^{n+s} \int_0^t \mathscr{R}\, \{e^{\Lambda_j(t-\tau)} \boldsymbol{q}_j \boldsymbol{q}_j'\}\, \boldsymbol{f}(\tau)\, d\tau.$$

As we will see, the case in which

$$\boldsymbol{f}(t) = \boldsymbol{f}_0 e^{i\omega t}, \tag{7.5.13}$$

where $\boldsymbol{f}_0$ is a constant vector and ω is real, is of considerable practical importance. If $i\omega$ is not a latent root of $D_2(\lambda)$ and Q is the $n \times 2n$ matrix of latent vectors, then we may use the solution (6.3.1) and obtain the particular integral:

$$\boldsymbol{p}(t) = e^{i\omega t} Q(I i\omega - \Lambda)^{-1} Q' \boldsymbol{f}_0.$$

Thus, writing $\boldsymbol{p} = \boldsymbol{q} e^{i\omega t}$, designating $\boldsymbol{q}$ as the *response vector*, and employing the partitions (7.5.9) and (7.5.10) we have†

$$\boldsymbol{q} = \{Q_1(Ii\omega - \Lambda_1)^{-1} Q_1' + Q_2(Ii\omega - \Lambda_2)^{-1} Q_2' + \bar{Q}_2(Ii\omega - \Lambda_2)^{-1} \bar{Q}_2'\} \boldsymbol{f}_0. \tag{7.5.14}$$

We also have the equivalent expressions:

$$\boldsymbol{q} = \sum_{j=1}^{2s} \frac{(\boldsymbol{q}_j' \boldsymbol{f}_0)}{i\omega - \lambda_j} \boldsymbol{q}_j + \sum_{j=2s+1}^{n+s} \left\{ \frac{\boldsymbol{q}_j \boldsymbol{q}_j'}{i\omega - \lambda_j} + \frac{\bar{\boldsymbol{q}}_j \bar{\boldsymbol{q}}_j'}{i\omega - \bar{\lambda}_j} \right\} \boldsymbol{f}_0, \tag{7.5.15}$$

$$= \sum_{j=1}^{2s} (\boldsymbol{q}_j' \boldsymbol{f}_0)\, \alpha_j e^{-i\theta_j} \boldsymbol{q}_j + \sum_{j=2s+1}^{n} (\boldsymbol{q}_j' \boldsymbol{f}_0)\, \beta_j e^{-i\varphi_j} \boldsymbol{q}_j + \sum_{j=2s+1}^{n+s} (\bar{\boldsymbol{q}}_j' \boldsymbol{f}_0)\, \gamma_j e^{-i\psi_j} \bar{\boldsymbol{q}}_j,$$

where (7.5.16)

$$\alpha_j = \{\omega^2 + \lambda_j^2\}^{-1/2}, \quad \theta_j = \tan^{-1}(\omega/-\lambda_j), \quad j = 1, 2, \ldots, 2s;$$

$$\left.\begin{aligned} \beta_j &= \{(\omega - \omega_j)^2 + \mu_j^2\}^{-1/2}, & \varphi_j &= \tan^{-1}\{(\omega - \omega_j)/-\mu_j\} \\ \gamma_j &= \{(\omega + \omega_j)^2 + \mu^2\}^{-1/2}, & \psi_j &= \tan^{-1}\{(\omega + \omega_j)/-\mu_j\} \end{aligned}\right\},$$

$$j = 2s+1, \ldots, n+s,$$

and $\lambda_j = \mu_j + i\omega_j$ for $j = 2s+1, \ldots, n+s$. It should be borne in mind that, by Theorem 7.1,

$$\lambda_j \leq 0, \quad j = 1, 2, \ldots, 2s,$$

and

$$\mu_j \leq 0, \quad j = 2s+1, \ldots, n+s.$$

† The identity matrices, I, appearing in this equation are to be interpreted as having the appropriate orders.

We see that the resonance phenomenon may still arise if there are complex latent roots whose real parts are very small. The coefficients β_j arising in the second summation of (7.5.16) may then be very large when ω is in some neighborhood of ω_j. Thus the response in the mode $\boldsymbol{q}_j$ may dominate that due to the other terms in (7.5.16), provided that $\boldsymbol{f}_0$ is not orthogonal (or nearly orthogonal) to the subspace of latent vectors of λ_j (cf. (7.4.9)).

Notice that these results still depend on the conditions (i) and (ii) stated above. Alternative results in which these conditions may be relaxed can be obtained from (6.6.2). However, stronger (but in practice not more restrictive) conditions corresponding to those of Theorem 3.5 must be satisfied. There appears to be little or no advantage in decompositions like those of (7.5.9) and (7.5.10) in the general case, though the problem considered in the next section provides an exception.

The problems considered in this section are very much simplified if the matrices A, B, and C are linearly dependent. For example, if there exist constants α, γ such that

$$B = \alpha A + \gamma C,$$

then the matrix Q which transforms A and C to diagonal matrices I and W^2 (Theorem 2.1) will transform B to a diagonal matrix also; for we then have

$$\begin{aligned} Q'BQ &= \alpha Q'AQ + \gamma Q'CQ = I\alpha + \gamma W^2 \\ &= \operatorname{diag}\{\alpha + \gamma\omega_1^2, \ldots, \alpha + \gamma\omega_n^2\}. \end{aligned}$$

In such cases (7.5.8) may immediately be reduced to n scalar equations of the form

$$\ddot{\xi} + k\dot{\xi} + \omega^2\xi = \boldsymbol{q}'\boldsymbol{f}(t),$$

for which k is a constant, and for which the solution is elementary (cf. (7.1.4)).

We have so far ignored the fact that in practice the coefficients of B are generally of smaller magnitude than those of A and C. If we make this assumption, useful approximate analyses can be made using the perturbation and other techniques. Some of these will be investigated in Chapter 9.

7.6 Overdamped Systems

We have seen that the mathematical analysis of vibrating systems with no frictional forces present is relatively easy from the mathematical point of view. We will see in Chapter 9 that by means of perturbation theory we can investigate lightly damped systems in terms of the properties of the undamped system. We now investigate cases in which the damping is heavy (in a sense to be defined) and it will transpire that in these two extreme cases the λ-matrix (7.5.2) under investigation is simple. In the general case considered in the previous section we *assumed* that (7.5.2) was a simple λ-matrix in order to keep the manipulations tractable. In the special cases of this and the following section we will *prove* this to be the case.

Let x be an arbitrary real vector of order n. We define the scalar a in terms of x by

$$a(x) = x'Ax$$

and similarly for $b(x)$, $c(x)$. The *overdamping condition* is now defined to be

$$b^2(x) - 4a(x)\,c(x) > 0 \tag{7.6.1}$$

for all $x \neq 0$ in $\mathscr{R}_n$. To summarize our hypotheses on an overdamped system we have:

(a) A, B, C are real and symmetric.
(b) A and B are positive definite $\big(a(x) > 0,\ b(x) > 0\big)$; C is non-negative definite $\big(c(x) \geqq 0\big)$.
(c) Inequality (7.6.1) is satisfied.

There is a temptation to continue with the assumption that B should be merely non-negative definite, but with this assumption and those on A and C, (7.6.1) then implies that B must be positive definite. For the remainder of this section we will refer consistently to the λ-matrix $D(\lambda) = A\lambda^2 + B\lambda + C$ in the overdamped case.

Theorem 7.2 *All the latent roots of $D(\lambda)$ are real and non-positive.*

The proof of this theorem is left to the reader. The ideas employed in proving Theorem 7.1 are used again. In particular, (7.5.7) is crucial.

Theorem 7.3 *$D(\lambda)$ is a simple λ-matrix.*

Proof Consider Theorem 4.6 in relation to $D(\lambda)$. We have $D^{(1)}(\lambda) = 2A\lambda + B$, and if the theorem were false there would exist a real latent root λ with (real) latent vector $\boldsymbol{q}$ such that

$$\boldsymbol{q}'(2A\lambda + B)\,\boldsymbol{q} = 0.$$

That is,

$$2a(\boldsymbol{q})\,\lambda + b(\boldsymbol{q}) = 0.$$

But we also have

$$(A\lambda^2 + B\lambda + C)\,\boldsymbol{q} = \boldsymbol{0},$$

which implies

$$a(\boldsymbol{q})\,\lambda^2 + b(\boldsymbol{q})\,\lambda + c(\boldsymbol{q}) = 0,$$

and hence

$$2a(\boldsymbol{q})\,\lambda + b(\boldsymbol{q}) = \pm\sqrt{b^2 - 4ac},$$

whereas the right-hand side is non-zero by the overdamping condition. Thus we arrive at a contradiction, and the theorem must be true. Let us recall that the multiplicity of a latent root is now equal to the dimension of the subspace of its latent vectors.

We define

$$d(\boldsymbol{x}) = \sqrt{b^2(\boldsymbol{x}) - 4a(\boldsymbol{x})\,c(\boldsymbol{x})}$$

and then

$$p_1(\boldsymbol{x}) = \frac{-b(\boldsymbol{x}) + d(\boldsymbol{x})}{2a(\boldsymbol{x})}, \qquad p_2(\boldsymbol{x}) = \frac{-b(\boldsymbol{x}) - d(\boldsymbol{x})}{2a(\boldsymbol{x})},$$

noting that $a(\boldsymbol{x}) \neq 0$ for any $\boldsymbol{x}$, since A is positive definite. We call $p_1(\boldsymbol{x})$ and $p_2(\boldsymbol{x})$ the *primary* and *secondary functionals* respectively. Clearly, for any $k \neq 0$,

$$p_i(k\boldsymbol{x}) = p_i(\boldsymbol{x}), \qquad i = 1, 2, \tag{7.6.2}$$

and, by definition,

$$\left.\begin{aligned} ap_1^2 + bp_1 + c &= 0, \\ 2ap_1 + b &> 0 \end{aligned}\right\} \tag{7.6.3}$$

and

$$\left.\begin{aligned} ap_2^2 + bp_2 + c &= 0, \\ 2ap_2 + b &< 0 \end{aligned}\right\} \tag{7.6.4}$$

for a, b, c (and hence p_1, p_2) defined at any $\boldsymbol{x} \neq \boldsymbol{0}$.

If $\boldsymbol{q}$ is a latent vector of $D(\lambda)$ we may now define $\boldsymbol{q}$ to be a *primary latent vector* or *secondary latent vector* according as the corresponding latent root is given by $p_1(\boldsymbol{q})$ or $p_2(\boldsymbol{q})$. We then have:

THEOREM 7.4 *The latent vectors of a given latent root are either all primary or all secondary latent vectors.*

Proof If the multiplicity of a latent root is one, the result follows immediately from (7.6.2). We therefore consider a multiple latent root λ which is supposed to have a primary latent vector $\boldsymbol{s}$ and a secondary latent vector $\boldsymbol{t}$. Let

$$\xi = (1 - \varrho)\,\boldsymbol{s} + \varrho\boldsymbol{t}.$$

For any real number ϱ, ξ is a latent vector with latent root λ and

$$a(\xi)\,\lambda^2 + b(\xi)\,\lambda + c(\xi) = 0,$$

which implies that

$$2a(\xi)\,\lambda + b(\xi) = \pm d(\xi).$$

Now $d(\xi)$ is a continuous function of ϱ and the overdamping condition ensures that it never vanishes. Hence $2a(\xi)\,\lambda + b(\xi)$ is of one sign for *all* ϱ. But our hypothesis implies that, at $\varrho = 0$ and $\varrho = 1$, $\lambda = p_1(\boldsymbol{s})$ and $\lambda = p_2(\boldsymbol{t})$ respectively, $2ap_1 + b > 0$ and $2ap_2 + b < 0$. Thus we arrive at a contradiction. The theorem is therefore proved.

The theorem now allows us to classify a latent root as primary or secondary according as all of its latent vectors are primary or secondary latent vectors.

LEMMA *Let λ_1, λ_2 be primary latent roots with latent vectors $\boldsymbol{x}_1$, $\boldsymbol{x}_2$ respectively and suppose that $\lambda_1 > \lambda_2$; then $\boldsymbol{x}_1$, $\boldsymbol{x}_2$ are linearly independent and, if $\boldsymbol{v}$ is any linear combination of $\boldsymbol{x}_1$ and $\boldsymbol{x}_2$, then*

$$\lambda_2 \leqq p_1(\boldsymbol{v}) \leqq \lambda_1.$$

Proof If $\boldsymbol{x}_1$ and $\boldsymbol{x}_2$ are linearly dependent, then one is a multiple of the other and equation (7.6.2) then implies that $p_1(\boldsymbol{x}_1) = p_1(\boldsymbol{x}_2)$. But $p_1(\boldsymbol{x}_1) = \lambda_1$ and $p_1(\boldsymbol{x}_2) = \lambda_2$ and $\lambda_1 \neq \lambda_2$. Hence $\boldsymbol{x}_1$ and $\boldsymbol{x}_2$ must be linearly independent.

Since the magnitude of $\boldsymbol{v}$ is not relevant in discussing $p_1(\boldsymbol{v})$, we may assume without loss of generality that

$$\boldsymbol{v} = (1 - \varrho)\,\boldsymbol{x}_1 + \varrho\boldsymbol{x}_2$$

for any real number ϱ. It is easily verified that

$$a(\boldsymbol{v})\,\lambda_1^2 + b(\boldsymbol{v})\,\lambda_1 + c(\boldsymbol{v}) = \varrho^2\,\{a(\boldsymbol{x}_2)\,\lambda_1^2 + b(\boldsymbol{x}_2)\,\lambda_1 + c(\boldsymbol{x}_2)\}.$$

Since λ_2 is the greatest zero of $a(\boldsymbol{x}_2)\,\lambda^2 + b(\boldsymbol{x}_2)\,\lambda + c(\boldsymbol{x}_2)$ and $\lambda_1 > \lambda_2$, it follows that the expression on the right of this equation is positive for all $\varrho \neq 0$. Thus,

$$a(\boldsymbol{v})\,\lambda_1^2 + b(\boldsymbol{v})\,\lambda_1 + c(\boldsymbol{v}) > 0,$$

which implies that *either*

or

$$\left.\begin{array}{l}\lambda_1 > p_1(\boldsymbol{v})\\ \lambda_1 < p_2(\boldsymbol{v})\end{array}\right\} \quad \text{for all } \varrho \neq 0.$$

Now at $\varrho = 1$, $p_1(\boldsymbol{v}) = p_1(\boldsymbol{x}_2) = \lambda_2 < \lambda_1$, so that the former alternative must be taken. Thus, for any ϱ whatever, $p_1(\boldsymbol{v}) \leqq \lambda_1$. A similar method of proof can be employed to show that $\lambda_2 \leqq p_1(\boldsymbol{v})$.

It should be noted that $p_1(\boldsymbol{v})$ depends continuously on ϱ so that every number between λ_1 and λ_2 belongs to the range of $p_1(\boldsymbol{v})$.

THEOREM 7.5 *Let $\lambda_1 > \lambda_2 > \cdots > \lambda_m$ be primary latent roots with latent vectors $\boldsymbol{x}_1, \boldsymbol{x}_2, \ldots, \boldsymbol{x}_m$ respectively; then these vectors are linearly independent and, if $\boldsymbol{z}$ is any linear combination of them, then*

$$\lambda_m \leqq p_1(\boldsymbol{z}) \leqq \lambda_1. \tag{7.6.5}$$

Proof The proof of the inequality is by induction. We note, first, that the case $m = 2$ has been proved in the lemma. The induction hypothesis, then, is that, if $\boldsymbol{w}$ is any linear combination of $\boldsymbol{x}_2, \ldots, \boldsymbol{x}_m$, then

$$\lambda_m \leqq p_1(\boldsymbol{w}) \leqq \lambda_2.$$

Let $\boldsymbol{z} = \varrho\boldsymbol{w} + (1 - \varrho)\,\boldsymbol{x}_1$; then it can be shown that

$$a(\boldsymbol{z})\,\lambda_1^2 + b(\boldsymbol{z})\,\lambda_1 + c(\boldsymbol{z}) = \varrho^2\{a(\boldsymbol{w})\,\lambda_1^2 + b(\boldsymbol{w})\,\lambda_1 + c(\boldsymbol{w})\}.$$

Now the greatest zero of $a(\boldsymbol{w})\,\lambda^2 + b(\boldsymbol{w})\,\lambda + c(\boldsymbol{w})$ is $p_1(\boldsymbol{w})$, which, by hypothesis, does not exceed λ_2. Since $\lambda_2 < \lambda_1$ and $a > 0$, we observe that the right (and therefore the left) side of the above equation is positive for all $\varrho \neq 0$. Thus (as in the lemma), either

or $$\left.\begin{matrix} \lambda_1 > p_1(\boldsymbol{z}) \\ \lambda_2 < p_2(\boldsymbol{z}) \end{matrix}\right\} \quad \text{for all } \varrho \neq 0.$$

Examination of the case $\varrho = 1$ shows that $\lambda_1 > p_1(\boldsymbol{z})$ for $\varrho \neq 0$ and when $\varrho = 0$ we get $\lambda_1 = p_1(\boldsymbol{z})$, so that (7.6.5) is proved and the induction is complete.

The independence of the latent vectors is established if we note that $p_1(\boldsymbol{w}) \neq p_1(\boldsymbol{x}_1)$ so that, by (7.6.2), $\boldsymbol{x}_1$ cannot be a linear combination of $\boldsymbol{x}_2, \boldsymbol{x}_3, \ldots, \boldsymbol{x}_m$.

If the word "primary" in Theorem 7.5 is replaced everywhere by "secondary", and p_1 is replaced by p_2, the resulting statement is also true. It would be repetitious to indicate the proofs in this case.

THEOREM 7.6 *If the latent roots are counted according to their multiplicities, there are n primary latent roots and n secondary latent roots. The primary latent vectors span $\mathscr{R}_n$ and so do the secondary latent vectors.*

Proof If there are k primary latent roots, then by the preceding theorem we can construct a set of k independent latent vectors in $\mathscr{R}_n$; hence $k \leqq n$. Similarly, if there are j secondary latent roots we have $j \leqq n$. But $j + k = 2n$, so that $j = k = n$. The final statements of the theorem are now obvious.

THEOREM 7.7 *Every primary latent root exceeds every secondary latent root.*

Proof Let λ_1, λ_n (with $\lambda_n \leqq \lambda_1$) be the extreme primary latent roots and $\lambda_{n+1}, \lambda_{2n}$ (with $\lambda_{2n} \leqq \lambda_{n+1}$) be the extreme secondary latent roots. We deduce from Theorems 7.5, 7.6 and the lemma to 7.5 that, for any vector $\boldsymbol{x}$, $\lambda_n \leqq p_1(\boldsymbol{x}) \leqq \lambda_1$ and $\lambda_{2n} \leqq p_2(\boldsymbol{x}) \leqq \lambda_{n+1}$. Also, for any λ such that $\lambda_n \leqq \lambda \leqq \lambda_1$ or $\lambda_{2n} \leqq \lambda \leqq \lambda_{n+1}$, there exists an $\boldsymbol{x}$ such that $p_1(\boldsymbol{x}) = \lambda$ in the former case or $p_2(\boldsymbol{x}) = \lambda$ in the latter. Furthermore, we have by definition that, for any $\boldsymbol{v}$, $p_1(\boldsymbol{v}) > p_2(\boldsymbol{v})$, so that the theorem will be established if we show that the range of values of p_2 includes no primary latent roots.

To this end, let λ be a primary latent root with latent vector $\boldsymbol{x}$ and let $\boldsymbol{y}$ be any non-zero vector for which

$$p_2(\boldsymbol{y}) = \lambda = p_1(\boldsymbol{x}).$$

Now $p_1(\boldsymbol{y}) \neq p_2(\boldsymbol{y})$, so that $p_1(\boldsymbol{y}) \neq p_1(\boldsymbol{x})$, and using (7.6.2) we deduce that $\boldsymbol{x}$ and $\boldsymbol{y}$ are linearly independent. We define

$$\boldsymbol{v} = \varrho \boldsymbol{y} + (1 - \varrho)\, \boldsymbol{x},$$

then

$$a(\boldsymbol{v})\, \lambda^2 + b(\boldsymbol{v})\, \lambda + c(\boldsymbol{v}) = \varrho^2 \{a(\boldsymbol{y})\, \lambda^2 + b(\boldsymbol{y})\, \lambda + c(\boldsymbol{y})\}.$$

However, $p_2(\boldsymbol{y}) = \lambda$ implies that the right side vanishes and

$$a(\boldsymbol{v})\, \lambda^2 + b(\boldsymbol{v})\, \lambda + c(\boldsymbol{v}) = 0$$

for *all* ϱ. Now for $\varrho = 1$, $2\lambda a + b < 0$ and for $\varrho = 0$, $2\lambda a + b > 0$. Furthermore, $2\lambda a + b$ is a continuous function of ϱ, and hence there exists a ϱ_0 such that $0 < \varrho_0 < 1$ and

$$2\lambda a(\boldsymbol{v}_0) + b(\boldsymbol{v}_0) = 0$$

where $\boldsymbol{v}_0 = \boldsymbol{v}(\varrho_0)$. But this implies $d(\boldsymbol{v}_0) = 0$, and hence $\boldsymbol{v}_0 = \boldsymbol{0}$, which contradicts the fact that $\boldsymbol{x}$ and $\boldsymbol{y}$ are independent. Hence there is no primary latent root in the range of p_2, and the theorem is proved.

An overdamped system is seen to have many properties which cannot be generalized to less heavily damped systems. Nevertheless, the overdamped case is of practical interest, and provides us with practical examples of simple λ-matrices. Furthermore, we see that the stronger conditions of Theorem 3.5 are satisfied. We may apply that result as follows: Let $\Lambda_1 = \operatorname{diag}\{\lambda_1, \ldots, \lambda_n\}$ be the diagonal matrix of primary latent roots and Q_1 be an $n \times n$ matrix whose columns are linearly independent primary latent vectors associated with $\lambda_1, \lambda_2, \ldots, \lambda_n$. Define Λ_2, Q_2 similarly in terms of the secondary roots and vectors. Then

$$\begin{aligned} A\lambda^2 + B\lambda + C &= \{I\lambda - (Q_2\Lambda_2Q_2^{-1})'\}\,A\,\{I\lambda - Q_1\Lambda_1Q_1^{-1}\} \\ &= (Q_2^{-1})'(I\lambda - \Lambda_2)\,Q_2'AQ_2(I\lambda - \Lambda_1)\,Q_1^{-1}. \end{aligned}$$

7.7 Gyroscopic Systems

The analysis of a conservative, holonomic mechanical system in terms of coordinates which are defined relative to a moving frame of reference gives rise to equations of motion which are rather more complicated than (7.3.1) or (7.4.2). The analysis of such systems is described by Whittaker [1] and by Frazer *et al.* [1]. The equations of motion are of the form

$$A\ddot{\boldsymbol{q}} + B\dot{\boldsymbol{q}} + C\boldsymbol{q} = \boldsymbol{0},$$

where

(a) A and C are real and symmetric.
(b) A is positive definite and C is non-negative definite.
(c) B is real and skew-symmetric.

Solutions of this differential equation may again be defined in terms of sets of latent values of

$$D(\lambda) = A\lambda^2 + B\lambda + C,$$

and we refer to the latent values of this particular problem for the remainder of this section.

Theorem 7.8 *All the latent roots are pure imaginary or zero.*

Proof Let λ be a latent root with right latent vector $\boldsymbol{q}$; then

$$(A\lambda^2 + B\lambda + C)\,\boldsymbol{q} = \boldsymbol{0}. \qquad (7.7.1)$$

Premultiply by $\bar{\boldsymbol{q}}'$, and write $a = \bar{\boldsymbol{q}}'A\boldsymbol{q}$, $ib = \bar{\boldsymbol{q}}'B\boldsymbol{q}$, and $c = \bar{\boldsymbol{q}}'C\boldsymbol{q}$, where $a > 0$, $c \geqq 0$, and b is real (cf. eqns. (1.5.6) and (1.5.7)). Then

$$a\lambda^2 + ib\lambda + c = 0.$$

Since $a > 0$ we write the solutions in the form

$$\lambda = i(-b \pm \sqrt{b^2 + 4ac}\,)/2a. \qquad (7.7.2)$$

reg. λ matrix 1. Square
2. M or A non-singular

Hence all the latent roots are of the form $i\omega$ where ω is real.

Equation (7.7.1) is now:

$$(-A\omega^2 + i\omega B + C)\,\boldsymbol{q} = \boldsymbol{0}.$$

Taking complex conjugates and transposing we then find that

$$\bar{\boldsymbol{q}}'(-A\omega^2 + i\omega B + C) = \boldsymbol{0}',$$

since $B' = -B$, $A' = A$, and $C' = C$. Thus with each right latent vector $\boldsymbol{q}$ there is associated a left latent vector $\bar{\boldsymbol{q}}$. Furthermore, if the subspace of right latent vectors of $i\omega$ is spanned by $\boldsymbol{q}_1, \boldsymbol{q}_2, \ldots, \boldsymbol{q}_\alpha$, then the subspace of left latent vectors of $i\omega$ is spanned by $\bar{\boldsymbol{q}}_1, \ldots, \bar{\boldsymbol{q}}_\alpha$. We are now in a position to prove:

THEOREM 7.9 *$D(\lambda)$ is a simple λ-matrix.*

Proof If the theorem is false, then by the above discussion and Theorem 4.6, there would exist a latent root $i\omega$ with right latent vector $\boldsymbol{q}$ such that

$$\bar{\boldsymbol{q}}'(2i\omega A + B)\boldsymbol{q} = 0,$$

that is, $2\omega a + b = 0$. But putting $\lambda = i\omega$ in (7.7.2) we also have

$$2\omega a + b = \pm\sqrt{b^2 + 4ac} \neq 0,$$

and we arrive at a contradiction. Hence the theorem is true.

The transformation of (7.7.1) to a problem in real (rather than complex) arithmetic is interesting. Put $\lambda = i\omega$ and $\boldsymbol{q} = \boldsymbol{u} + i\boldsymbol{v}$ where ω, $\boldsymbol{u}$, and $\boldsymbol{v}$ are real. Then equate real and imaginary parts of the left-hand side to zero and the resulting equations can be written in the form:

$$\left\{\omega^2\begin{bmatrix} A & 0 \\ 0 & A \end{bmatrix} + \omega\begin{bmatrix} 0 & B \\ -B & 0 \end{bmatrix} - \begin{bmatrix} C & 0 \\ 0 & C \end{bmatrix}\right\}\begin{bmatrix} \boldsymbol{u} \\ \boldsymbol{v} \end{bmatrix} = \boldsymbol{0}.$$

Now $B' = -B$, so the coefficient of ω in this equation is real and symmetric. Furthermore, it is easily verified that the $2n \times 2n$ coefficient matrices here satisfy the overdamping condition (7.6.1), and it follows that every latent root ω is real, as we would expect. In fact, the spectrum of latent roots of (7.7.1) is reproduced twice in this real formulation. From the computational point of view this property is often detrimental, for many iterative processes converge more rapidly to simple than to multiple roots (cf. Chapter 5).

Simple λ-matrix 1. Reg.
2. no degeneracy

7.8 Sinusoidal Motion with Hysteretic Damping

While considering particular integrals under the action of sinusoidal applied forces of the form (7.5.13), we may also consider the effect of hysteretic damping, as described in § 7.2. It is important to remember that (7.2.8) and (7.2.9) are only applicable in sinusoidal motion. Thus, when we employ the equations of motion (7.2.10) in the form

$$A\ddot{\boldsymbol{p}} + B\dot{\boldsymbol{p}} + (C + iG)\,\boldsymbol{p} = \boldsymbol{f}_0 e^{i\omega t}, \tag{7.8.1}$$

we ignore any complementary functions, since there are no grounds for asserting that they will be sinusoidal. However, as we will see, particular solutions do exist which represent sinusoidal motion.

The only admissible solutions of (7.8.1) are now of the form

$$\boldsymbol{p}(t) = \boldsymbol{q}e^{i\omega t}, \tag{7.8.2}$$

where $\boldsymbol{q}$ may be complex and ω must be real. For such solutions we have

$$(-A\omega^2 + iB\omega + C + iG)\,\boldsymbol{q} = \boldsymbol{f}_0. \tag{7.8.3}$$

Or, writing $\lambda = i\omega$, we obtain

$$(A\lambda^2 + B\lambda + C + iG)\,\boldsymbol{q} = \boldsymbol{f}_0. \tag{7.8.4}$$

Once more the problem is reduced to that of inverting a λ-matrix of second degree. The matrix is now complex, but this does not affect the validity of the spectral forms for the inverse given in Theorem 4.3 and in (4.6.1). The latent roots and vectors will in general be complex (and not in conjugate pairs) and they will be $2n$ in number. However, the λ-matrix is symmetric so that the left and right latent vectors coincide.

We repeat that, unlike the problem in which $G = 0$, the latent roots and vectors of

$$A\lambda^2 + B\lambda + C + iG \tag{7.8.5}$$

have no physical significance. Apart from the fact that (7.2.8) was formulated for sinusoidal motion, the difficulty is that, if $\boldsymbol{q}$ and λ are associated complex latent values, then $\boldsymbol{q}e^{\lambda t}$ is a solution of

$$A\ddot{\boldsymbol{p}} + B\dot{\boldsymbol{p}} + (C + iG)\,\boldsymbol{p} = \boldsymbol{0},$$

but $\mathscr{R}(\boldsymbol{q}e^{\lambda t})$ and $\mathscr{I}(\boldsymbol{q}e^{\lambda t})$ are not solutions of this equation. Hence, the concept of a complex notation, as introduced in § 7.1 breaks down.

In view of the fact that the latent roots may yet be important in the application of Theorem 4.3 we state the following properties, all of which are easily deduced by the method of proof of Theorem 7.1.

(i) If null latent roots exist, then both G and C are non-negative definite (though the converse is not necessarily true).
(ii) If real non-zero latent roots exist, they are all negative and G must be non-negative definite.
(iii) If G is positive definite, then all the latent roots are complex.
(iv) If $B = 0$, then all the complex latent roots arise in pairs of the form $\pm(\mu + i\gamma)$ where μ and γ are real and are of opposite sign.

Once more, useful approximate methods can be employed if, as is usually the case, B and G are of small magnitude when compared with A and C.

7.9 Solutions for Some Non-conservative Systems

We now consider briefly those problems which give rise to equations of motion of the form

$$A_0\ddot{\boldsymbol{p}} + A_1\dot{\boldsymbol{p}} + A_2\boldsymbol{p} = \boldsymbol{0}, \tag{7.9.1}$$

in which A_0, A_1, and A_2 are all real but not necessarily symmetric. Equations of this kind arise in the treatment of aircraft flutter problems. In this case the equations of motion (7.2.11) are formulated, although the generalized (aerodynamic) forces are contained in the vector $\boldsymbol{P}$, but are themselves linear functions of the displacement vector $\boldsymbol{p}$ and its time derivatives. This gives rise to a classical example of *self-excited* oscillations, in which all the external forces are zero if the system is at rest in its equilibrium position. We now suppose that $\boldsymbol{P}$ depends only on $\boldsymbol{p}$, $\dot{\boldsymbol{p}}$, and $\ddot{\boldsymbol{p}}$, and if the dependence is linear, equations of the form (7.9.1) are obtained. In flutter theory A_0, A_1, and A_2 are defined in terms of a real parameter, usually the airspeed (re. § 5.9).

For the solutions of (7.9.1) we can hardly improve on those presented in Chapter 6, which we will now briefly discuss. As we saw in § 6.2, all the solutions of (7.9.1) can be expressed as linear combinations of the fundamental solutions

$$\boldsymbol{p}(t) = \boldsymbol{q}e^{\lambda t}, \tag{7.9.2}$$

where λ is a latent root of

$$D_2(\lambda) = A_0\lambda^2 + A_1\lambda + A_2, \tag{7.9.3}$$

$\boldsymbol{q}$ is a corresponding right latent vector, and provided $D_2(\lambda)$ is simple. As in § 7.5 we define the stability of these solutions by the sign of $\mathscr{R}(\lambda)$. The latent roots are real or occur in conjugate pairs, so that partitions of the form (7.5.9) and (7.5.10) could be employed, if this was felt to be desirable.

For the non-homogeneous equations

$$A_0\ddot{\boldsymbol{p}} + A_1\dot{\boldsymbol{p}} + A_2\boldsymbol{p} = \boldsymbol{f}(t) \tag{7.9.4}$$

we obtain the complete solution from (6.2.13) and (6.2.14). In this case we assume once more that $D_2(\lambda)$ is a simple λ-matrix and that the conditions of Theorem 4.5 are satisfied. We find that

$$\begin{aligned}\mathscr{D}^r\,\boldsymbol{p}(t) &= Q\Lambda^r e^{\Lambda t}\{\boldsymbol{c} + \boldsymbol{\varphi}(t)\}, \qquad r = 0, 1, \\ &= \sum_{j=1}^{2n} \lambda_j^r e^{\lambda_j t}\{c_j + \varphi_j(t)\}\boldsymbol{q}_j,\end{aligned} \tag{7.9.5}$$

where $\mathscr{D} = d/dt$, $\boldsymbol{c}$ is a column vector of $2n$ constants, and $\boldsymbol{\varphi}(t)$ has elements

$$\varphi_j(t) = \int_0^t e^{-\lambda_j\tau}\boldsymbol{r}_j'\boldsymbol{f}(\tau)\,d\tau. \tag{7.9.6}$$

Alternatively, if the conditions of Theorem 3.5 are satisfied, then, without normalizing the latent vectors or constructing biorthogonal systems (should this be necessary), we obtain the following particular integral from (6.6.2):

$$\boldsymbol{p}(t) = Q_1\int_0^t \{\Omega e^{\Lambda_2(t-\nu)} - e^{\Lambda_1(t-\nu)}\Omega\}R_2'\boldsymbol{f}(\nu)\,d\nu, \tag{7.9.7}$$

where

$$\Omega = \left\{\frac{t_{jk}}{\lambda_{n+k} - \lambda_j}\right\}, \qquad j, k = 1, 2, \ldots, n,$$

and $T = (R_2'A_0Q_1)^{-1}$.

Finally, if the forcing function $\boldsymbol{f}$ depends exponentially on t, then corresponding results may be obtained from (6.3.2). In particular if $\boldsymbol{f}(t) = \boldsymbol{f}_0 e^{i\omega t}$ as defined in (7.5.13), then the solution (7.9.5) gives

$$\boldsymbol{p}(t) = \boldsymbol{q}e^{i\omega t}$$

and

$$\boldsymbol{q} = \sum_{j=1}^{2n} \frac{\boldsymbol{r}_j' \boldsymbol{f}_0}{i\omega - \lambda_j} \boldsymbol{q}_j, \tag{7.9.8}$$

provided $i\omega$ is not a latent root of $D_2(\lambda)$. The resonance phenomenon may still arise and is clearly illustrated in this equation, as it was in (7.5.15) for the symmetric case. Thus, if λ_j is a latent root whose real part is very small (i.e., corresponds to a lightly damped mode), then in some neighborhood of $\omega = \mathscr{I}(\lambda_j)$ a response vector which is of large magnitude, and which lies in the subspace of right latent vectors of λ_j, is obtained, provided that $\boldsymbol{f}_0$ is not orthogonal to the subspace of *left* latent vectors of λ_j.

We will see in the next section that the vectors $\boldsymbol{q}_j$ define the free modes of vibration for the system, and (7.9.8) shows how every response vector $\boldsymbol{q}$ may be expressed as a linear combination of these modes. The left latent vectors $\boldsymbol{r}_j$ have no obvious physical significance, and yet they play a very important role in determining the relative amplitudes of the free modes in the response vector.

The result (7.9.8) should be particularly useful in the analysis of the response of a system to a load distribution which is random in time. The generalized forces may be defined by their power spectra, which resolve the mean square amplitudes of the forces into their frequency components (cf. Crandall [2] and Diederich [1]). Equations (7.9.8) can then be used to find the power spectra for the response.

7.10 Some Properties of the Latent Vectors

In § 7.5 and 7.9 we asserted that the right latent vectors $\boldsymbol{q}_j$ of (7.5.2) and (7.9.3) defined *free modes of vibration* for the system. We now justify this terminology by demonstrating that motion may occur in any one mode of free vibration without tending to disturb the system in the other such modes. We will restrict our attention to eqns. (7.9.1), as these include (7.5.1) as a special case. The solution corresponding to a right latent vector $\boldsymbol{q}_j$ with latent root λ_j is $\boldsymbol{p}(t) = \boldsymbol{q}_j e^{\lambda_j t}$, but, as we explained in § 7.1, it is the real part of this vector which defines the physical displacement of the system at time t.

Now suppose that the system is brought to rest and held in a position defined by $\mathscr{R}(\boldsymbol{q}_j)$. The system is then set in motion by a set of impulses which give rise to a velocity distribution defined by $\mathscr{R}(\lambda_j \boldsymbol{q}_j)$. These one-point boundary conditions (cf. § 6.4) are to define the $2n \times 1$ column vector $\boldsymbol{c}$ of the complete solution

$$\boldsymbol{p} = Qe^{\Lambda t}\boldsymbol{c}. \tag{7.10.1}$$

We also have

$$\dot{\boldsymbol{p}} = Q\Lambda e^{\Lambda t}\boldsymbol{c},$$

so, putting $t = 0$, the boundary conditions give

$$\begin{bmatrix} Q\Lambda \\ Q \end{bmatrix} \boldsymbol{c} = \begin{bmatrix} \lambda_j \boldsymbol{q}_j \\ \boldsymbol{q}_j \end{bmatrix},$$

which has the obvious unique solution in which $\boldsymbol{c}$ has a 1 in the jth position and zeros elsewhere. On substituting this vector in (7.10.1) we find that for all subsequent t the displacement is defined by

$$\boldsymbol{p}(t) = \boldsymbol{q}_j e^{\lambda_j t},$$

i.e., a motion involving only one latent root and vector.

Finally, we apply some of the results of § 4.4 and show how the coefficient matrices A_0, A_1, and A_2 may be expressed in terms of the latent roots and vectors. Once more we assume that the λ-matrix $D_2(\lambda)$ of (7.9.3) is a simple λ-matrix, so that the conditions of Theorem 4.5 may be satisfied. When this is the case we define the $n \times n$ matrices $\Gamma_r = Q\Lambda^r R'$, $r = \ldots -1, 0, 1, 2, \ldots$, where they exist, and obtain

$$0 = \Gamma_0,$$

$$A_0^{-1} = \Gamma_1, \tag{7.10.2}$$

$$A_1 = -A_0 \Gamma_2 A_0, \tag{7.10.3}$$

$$A_2 = A_0(\Gamma_2 A_0 \Gamma_2 - \Gamma_3) A_0. \tag{7.10.4}$$

These results are obtained from (4.4.8), (4.4.9), and (4.4.12). If it is known that A_2 is non-singular, then it follows from (4.4.10) that the last relation can be replaced by

$$A_2^{-1} = -\Gamma_{-1}. \tag{7.10.5}$$

Notice also that we may write

$$\Gamma_r = \sum_{j=1}^{2n} \lambda_j^r (\boldsymbol{q}_j \boldsymbol{r}_j').$$

Equations (7.10.2)–(7.10.5) appear to offer the possibility of *designing* a system which will have prescribed vibration properties.

That is, in designing a mechanical system, we may first assign the required natural frequencies and rates of decay, thus determining Λ. Then the related modes of vibration are assigned, determining Q and R; then we can compute the Γ_r, and hence A_0, A_1, and A_2. From these matrices we then obtain the distributions of structural properties which conform with the assigned vibration properties.

CHAPTER 8

ON THE THEORY OF RESONANCE TESTING

8.1 Introduction

Many engineering structures have elastic properties—mostly by accident rather than design. That is to say, the engineer would often be very well pleased if his designs could be made up in materials which are infinitely rigid. However, since elastic materials usually cannot be avoided, engineering structures usually distort from the shape in which they were designed when loads are applied to them. The effects of such distortions can be detrimental to the structure, especially if they vary rapidly with time, when vibration phenomena are likely to arise.

In order to predict the behavior of elastic structures under the influence of varying applied forces we usually idealize the system and replace it by a mathematical model with only a finite number of degrees of freedom. In many applications the most convenient set of coordinates to use in the model is a finite number of the *principal coordinates* for the undamped system, as defined in § 7.3. The related natural modes of vibration can usually be found by theoretical means, but it is also necessary to confirm the calculations by performing practical tests designed to measure these modes and the related natural frequencies.

These tests are often carried out by applying known sinusoidal forces to the structure, which is assumed to absorb only very small amounts of energy through damping effects. The frequency of the applied force is varied continuously, and the amplitude of the response in the structure is measured. It is then hoped that the approximate natural frequencies and modes may be determined by analysis of the resonance phenomenon described theoretically in § 7.5.

There are several obvious difficulties in this process. This is especially true if we try to analyze the problem by means of the result (7.5.16), for it appears that one may never be able to get a response related to a single vector $\boldsymbol{q}_j$, which would then allow a free mode of vibration for the damped system to be found. And even if we could obtain such a mode we would then want to deduce from it a normal mode of vibration for the *undamped* system. Here it seems intuitively reasonable to assume that a small cause—introducing the damping forces—will have a small effect on the modes of free vibration, but even this is not valid, as we shall see in Chapter 9. In any case, there will be considerable difficulties if any of the natural frequencies are nearly equal, for then the resonant peaks in the amplitude of the response will be superimposed on each other and be correspondingly difficult to analyze.

The problem appears to be a formidable one and might even, from a mathematical point of view, be thought to be insoluble. However, Fraeijs de Veubeke [1] has shown that, if the applied forces and forcing frequency are properly chosen and applied to a *damped* system (even a heavily damped system), the response may yet be in an *undamped* natural mode of vibration. In this chapter, it is our aim to put this elegant and apparently useful theory on a sound mathematical footing. It is not our aim to give an account of all the implications and limitations of the theory, but merely to expound those fundamental ideas which are amenable to an exact formulation within the framework of a finite-dimensional system.

8.2 The Method of Stationary Phase

In this chapter we will be concerned with particular integrals of (7.5.8) in which the right-hand side is defined by (7.5.13). We may assume that particular integrals are all sinusoidal with frequency ω, so the concept of hysteretic damping may also be employed, in which case we have, as in (7.2.10):

$$A\ddot{\boldsymbol{p}} + B\dot{\boldsymbol{p}} + (C + iG)\,\boldsymbol{p} = \boldsymbol{f}_0 e^{i\omega t}. \tag{8.2.1}$$

We now ask, can a *real* force amplitude vector $\boldsymbol{h}$ be found such that the elements of the response vector are in phase with one another (as in an undamped natural mode; cf. § 7.3), but not necessarily in phase with the elements of $\boldsymbol{h}$? That is, such that

$$\boldsymbol{p}(t) = \boldsymbol{v}e^{i(\omega t - \theta)}, \tag{8.2.2}$$

where $\boldsymbol{v}$ and θ are real? We assume that

$$0 \leqq \omega < \infty \qquad \text{and} \qquad 0 \leqq \theta < \pi. \tag{8.2.3}$$

If such solutions are possible, then (8.2.1) gives

$$\{(C - A\omega^2) + i(\omega B + G)\}\, \boldsymbol{v}e^{-i\theta} = \boldsymbol{h},$$

and equating real and imaginary parts we obtain

$$\text{and} \quad \left.\begin{aligned} \{(C - A\omega^2)\cos\theta + (\omega B + G)\sin\theta\}\, \boldsymbol{v} &= \boldsymbol{h} \\ \{(C - A\omega^2)\sin\theta - (\omega B + G)\cos\theta\}\, \boldsymbol{v} &= \boldsymbol{0}. \end{aligned}\right\} \tag{8.2.4}$$

Now let $t = \tan\theta$, and the latter equation becomes

$$\{(C - A\omega^2)\, t - (\omega B + G)\}\, \boldsymbol{v} = \boldsymbol{0} \tag{8.2.5}$$

and we view this as a latent root problem for t involving a symmetric matrix pencil whose coefficients are functions of ω.

Let ω_s denote a typical natural frequency, i.e., a zero of $|C - A\omega^2|$. Then, if $\omega \neq \omega_s$, $C - A\omega^2$ is non-singular, and $\omega B + G$ may be assumed to be non-negative definite. If follows from the theorem proved in the appendix to this chapter that when $\omega \neq \omega_s$ there exist n finite and *real* latent roots $t_1, t_2, \ldots, t_n$. There will therefore be at least one latent vector, and this will also be real. Each of the latent vectors with its associated latent root will define a solution (8.2.2) of the required form.

Now the maximal number of linearly independent latent vectors at any fixed value of ω is equal to n if and only if the pencil is a

$$(C - A\omega^2)\, t - (\omega B + G) \tag{8.2.6}$$

simple pencil at that value of ω (cf. § 2.1). If $\omega \neq \omega_s$ and $\omega B + G$ is positive definite, this is easily shown to be the case by means of Theorem 2.11. If $\omega B + G$ is merely non-negative definite we may at best *assume* that the pencil will, in general, be simple. An example in which this is not the case is also included in the appendix to this chapter. On the above assumption we define a non-singular matrix of independent latent vectors:

$$V(\omega) = [\boldsymbol{v}_1, \boldsymbol{v}_2, \ldots, \boldsymbol{v}_n],$$

and also

$$T = \operatorname{diag}\{t_1, t_2, \ldots, t_n\}.$$

We will call the vectors $\boldsymbol{v}_r(\omega)$ *proper vectors* and the latent roots $t_r(\omega)$ the *proper numbers*.

We now deduce from Theorem 2.1 that when $\omega \neq \omega_s$ the proper vectors may generally be defined in such a way that

$$V'(C - A\omega^2)\,V = D$$

and (8.2.7)

$$V'(\omega B + G)\,V = DT,$$

where D is a non-singular diagonal matrix which we may assign as we please. Now $\omega B + G$ is at least non-negative definite so that the diagonal entries of DT must be non-negative if we want V to be real. Thus, if $D = \text{diag}\,\{d_1, d_2, \ldots, d_n\}$ then $\text{sgn}\ d_r = \text{sgn}\ t_r$, $r = 1, 2, \ldots, n$. In fact, it is expedient to assume that the proper vectors are normalized so that

$$d_r = \cos\theta_r = \frac{\text{sgn}\ t_r}{(1 + t_r^2)^{1/2}}\,.$$

If $\omega \neq \omega_s$ we cannot have $\cos\theta_r = 0$; otherwise the second of eqns. (8.2.4) would provide a contradiction. If we write

$$P = \text{diag}\,\{\text{sgn}\ t_1, \ldots, \text{sgn}\ t_n\},$$

then we have

$$D = P(I + T^2)^{-1/2}, \tag{8.2.8}$$

and note that $P^{-1} = P$, $P^2 = I$.

We now show how a sinusoidal solution of (8.2.1) may be expressed in terms of the proper vectors and the spectrum of proper numbers. We assume always that (8.2.6) is a simple pencil. We first write $\boldsymbol{p}(t) = \boldsymbol{q}e^{i\omega t}$, where $\boldsymbol{q}$ is a complex vector independent of t, but dependent on ω. Then (8.2.1) gives

$$\{(C - A\omega^2) + i(\omega B + G)\}\,\boldsymbol{q} = \boldsymbol{f}_0.$$

We may substitute for $(C - A\omega^2)$ and $(\omega B + G)$ from (8.2.7) and (8.2.8) to obtain

$$(V')^{-1}\,P(I + T^2)^{-1/2}\,(I + iT)\,V^{-1}\boldsymbol{q} = \boldsymbol{f}_0. \tag{8.2.9}$$

Hence

$$\boldsymbol{q} = V(I + iT)^{-1}\,(I + T^2)^{1/2}\,PV'\boldsymbol{f}_0 \tag{8.2.10}$$

$$= V(I - iT)\,(I + T^2)^{-1/2}\,PV'\boldsymbol{f}_0. \tag{8.2.11}$$

In summation form:

$$\boldsymbol{q} = \sum_{r=1}^{n} (\text{sgn}\ t_r)\,\frac{1 - i\,t_r}{(1 + t_r^2)^{1/2}}\,(\boldsymbol{v}_r'\boldsymbol{f}_0)\boldsymbol{v}_r\,;$$

and putting $t_r = \tan\theta_r$ and using (8.2.3) we get

$$\boldsymbol{q} = \sum_{r=1}^{n} e^{-i\theta_r}(\boldsymbol{v}_r'\boldsymbol{f}_0)\,\boldsymbol{v}_r. \tag{8.2.12}$$

This result may be compared with equation (7.5.16) where the response is expressed in terms of the free modes of vibration and the (generally) complex latent roots of $D_2(\lambda)$. Though the latter expression seems more formidable, yet it is defined in terms of vectors and latent roots which are independent of ω, which is not the case for the more elegant result (8.2.12). In some applications the results of Chapter 7 may therefore be more appropriate, but it seems to be very difficult to utilize them in designing and analyzing practical resonance test techniques.

The matrix product appearing on the right of (8.2.11) is known as the *resolvent* or *receptance* matrix. We deduce from (8.2.12) that it may also be written in the form:

$$\sum_{r=1}^{n} e^{-i\theta_r}\boldsymbol{v}_r\boldsymbol{v}_r'.$$

Finally, we investigate the force vector $\boldsymbol{h}$ of (8.2.4) which gives rise to a response defined by a proper vector. In fact, let $\boldsymbol{h}_r$ be the force vector giving rise to a response determined by $\boldsymbol{v}_r$ and write

$$\sec\theta_r = (\operatorname{sgn} t_r)(1 + t_r^2)^{1/2};$$

then (8.2.4) gives

$$\{(C - A\omega^2) + (\omega B + G)\,t_r\}\,\boldsymbol{v}_r = (\operatorname{sgn} t_r)(1 + t_r^2)^{1/2}\,\boldsymbol{h}_r,$$

for $r = 1, 2, \ldots, n$. Thus

$$(C - A\omega^2)\,V + (\omega B + G)\,VT = HP(I + T^2)^{1/2},$$

where

$$H = [\boldsymbol{h}_1, \boldsymbol{h}_2, \ldots, \boldsymbol{h}_n].$$

Premultiplying the last result by V' and using (8.2.7, 8) we get

$$P(I + T^2)^{1/2} = V'HP(I + T^2)^{1/2},$$

and hence

$$V'H = I. \tag{8.2.13}$$

Thus, if $\omega \neq \omega_s$, and we have a complete set of proper vectors, then the *proper force vectors*, $\boldsymbol{h}_r(\omega)$, are linearly independent and form a biorthogonal system with the proper vectors. Our motive for normalizing in the apparently clumsy way leading to (8.2.8) is now apparent.

8.3 Properties of the Proper Numbers and Vectors

We have already noted that, if $\omega B + G$ is positive definite and $\omega \neq \omega_s$, then we are assured that the pencil (8.2.6) is a simple pencil. We will now formulate three theorems applicable to this case, and we will indicate later on how the results are modified if $\omega B + G$ is merely non-negative definite. We continue to denote a typical natural frequency for the undamped system by ω_s.

Theorem 8.1 *If $\omega B + G$ is positive definite, $\omega > 0$, and $\omega \neq \omega_s$, then there exist n finite, real, non-zero proper numbers and n linearly independent proper vectors.*

This result has already been used and, as we noted then, it follows immediately from Theorem 2.11.

Theorem 8.2 *If $\omega B + G$ is positive definite and $\omega = \omega_s$ where ω_s has multiplicity α, then there exist α infinite proper numbers and $n - \alpha$ finite, real, and non-zero proper numbers. In addition, there exists a complete set of real proper vectors, and those vectors corresponding to the infinite proper number span the subspace of the principal vectors (defining the natural modes of vibration) for $\omega = \omega_s$.*

These results are easily obtained by putting $t = 1/\mu$ in (8.2.5) to obtain a simple pencil in μ, even at $\omega = \omega_s$. The proper vectors associated with $\mu = 0$ are obviously the solutions of

$$(C - A\omega_s^2)\,\boldsymbol{x} = \boldsymbol{0},$$

and the solutions of this equation are the principal vectors of the natural frequency ω_s (cf. § 7.3).

Theorem 8.3 *If $\omega B + G$ is positive definite, then all the proper numbers are monotonic increasing functions of ω, except perhaps at $\omega = \omega_s$.*

Proof Let t, $\boldsymbol{v}$ be a proper number and vector respectively, such that

$$\{(C - A\omega^2)\,t - (\omega B + G)\}\,\boldsymbol{v} = \boldsymbol{0}. \tag{8.3.1}$$

Premultiply this equation by $\boldsymbol{v}'$ and write $\boldsymbol{v}'C\boldsymbol{v} = c$, etc. Then

$$(c - a\omega^2)\,t - (\omega b + g) = 0, \tag{8.3.2}$$

where

$$a > 0, \qquad b, c, g \geqq 0, \qquad \text{and} \qquad \omega b + g > 0. \tag{8.3.3}$$

When $\omega \neq \omega_s$ we may premultiply (8.3.1) by $(C - A\omega^2)^{-1}$ and observe that the proper numbers are simply the eigenvalues of the matrix

$$K = (C - A\omega^2)^{-1}(\omega B + G).$$

Furthermore, K has simple structure and the elements are rational functions of ω which are regular in some neighborhood of ω for $\omega \neq \omega_s$. We may therefore take it that the eigenvalue $t(\omega)$ and eigenvector $\boldsymbol{v}(\omega)$ are differentiable (cf. Theorem 2.5 and preceding remarks). Differentiating (8.3.1) with respect to ω and then premultiplying by $\boldsymbol{v}'$ we get

$$-2a\omega t + (c - a\omega^2)\frac{dt}{d\omega} - b = 0.$$

Hence

$$(c - a\omega^2)\, t\frac{dt}{d\omega} = bt + 2a\omega t^2,$$

and using (8.3.2):

$$(\omega b + g)\frac{dt}{d\omega} = bt + 2a\omega t^2. \tag{8.3.4}$$

Using (8.3.2) again we also have

$$(\omega b + g)\frac{dt}{d\omega} = \frac{1}{\omega}ct^2 + \omega a t^2 - gt.$$

We deduce from this equation and from (8.3.3) that $dt/d\omega > 0$ for all $t < 0$, and from (8.3.4) we have $dt/d\omega > 0$ for all $t > 0$. By Theorems 8.1 and 8.2 $t \neq 0$, and hence Theorem 8.3 is proved.

Clearly, the monotonicity will hold at $\omega = \omega_s$ for those proper numbers which remain finite as $\omega \to \omega_s$.

Thus far, the analysis has not been unduly complicated by the presence of both B and G, and in practical calculations there will usually be just one or the other present. In order to avoid considering too many subcases we will now assume that $G = 0$, and focus our attention on systems with viscous damping only. Similar results are obtained if only hysteretic damping is present and can very easily be formulated by the methods we are about to employ.

When $G = 0$ and $\omega \to 0$ the solutions of (8.3.1) approach those of

$$(Ct - \omega B)\,\boldsymbol{v} = \boldsymbol{0}. \tag{8.3.5}$$

Let r_b, r_c be the ranks of B and C; then if we are to use the three theorems above we must have $r_b = n$. In this case it is easily deduced that the pencil $Ct - \omega B$ has $n - r_c$ infinite latent roots and r_c positive latent roots which are proportional to ω. These define the behavior of the proper numbers as $\omega \to 0_+$.

When $\omega \to \infty$ the solutions of (8.3.1) approach those of

$$(A\omega t + B)\,\boldsymbol{v} = \boldsymbol{0}. \tag{8.3.6}$$

The pencil $A\omega t + B$ is simple and has n real and negative latent roots which are inversely proportional to ω. Thus, all the proper numbers of (8.3.1) tend to zero from below as $\omega \to \infty$.

The complete picture of the behavior of the proper numbers as functions of ω can now be built up. At $\omega = 0$ the curves $t_r(\omega)$ in the ω, t plane "start" from (0, 0) or from $(0, -\infty)$ and increase monotonically with ω. Each of the r_c curves starting from the origin has an asymptote at some ω_s when it changes sign from positive to negative as ω increases through ω_s. Those curves starting from $(0, -\infty)$ have the two coordinate axes as asymptotes. We note that no curve ever crosses the axis $t = 0$.

If the damping matrix is only non-negative definite, then the above results are modified considerably, even if we take the view that non-simple, or defective, pencils are only of pathological interest. For example, there may be a vector $\boldsymbol{x}$ which belongs to the null-spaces of both B and C, in which case $|Ct - \omega B| \equiv 0$ in (8.3.5) and the limiting behavior as $\omega \to 0$ cannot be obtained entirely from that equation. We should note that in this case the pencil $Ct - \omega B$ has less than n latent roots. However, if $|Ct - \omega B| \not\equiv 0$, it follows that

$$r_b + r_c > n,$$

and the limiting behavior as $\omega \to 0$ may still be obtained from (8.3.5). It is found that for $\omega \neq 0$ there exist $r_b + r_c - n$ latent roots of (8.3.5) which are proportional to ω, $n - r_b$ latent roots (and proper numbers of (8.3.1)) which are identically zero, and $n - r_c$ roots which are large and negative as $\omega \to 0_+$. For $\omega \neq 0$ or ω_s, there exist r_b proper numbers which are finite and not identically zero, and as $\omega \to \infty$, these proper numbers approach zero through negative values. If we observe that, for any proper vector $\boldsymbol{v}$ associated with a non-zero proper number t, $b \neq 0$ in (8.3.2), then the proof of Theorem 8.3 can be used again to show that $dt/d\omega > 0$ if $t \neq 0$ and t is finite.

Now the number of changes in sign of the functions $t_r(\omega)$ from positive to negative (occurring at the points $\omega = \omega_s$) is r_c, and this exceeds the number of initially positive numbers by $n - r_b$. Hence there exist $n - r_b$ points (not necessarily distinct) at which one of the curves $t_r(\omega)$ changes sign from negative to positive. By the monotone property this can only occur at zeros of the curves $t_r(\omega)$.

We have shown that there are, in general, $n - r_b$ zeros of the functions $t_r(\omega)$ at real, positive values of ω.

In Figs. 8.1 and 8.2 we give two very simple examples to illustrate some of the points we have made. In both cases

$$A = \begin{bmatrix} 1 & 0 \\ 0 & 1 \end{bmatrix} \quad \text{and} \quad C = \begin{bmatrix} 1 & \frac{1}{2} \\ \frac{1}{2} & 1 \end{bmatrix},$$

so that both A and C are positive definite and the natural frequencies are given by $\omega_1^2 = \frac{1}{2}, \omega_2^2 = \frac{3}{2}$. In the first example we have the unit matrix for the viscous damping coefficients, and in the second example

$$B = \begin{bmatrix} 1 & 0 \\ 0 & 0 \end{bmatrix},$$

which is non-negative definite and has rank one. In this case (Fig. 8.2) the curve $t(\omega)$ has a zero, and there is one proper number which is identically zero.

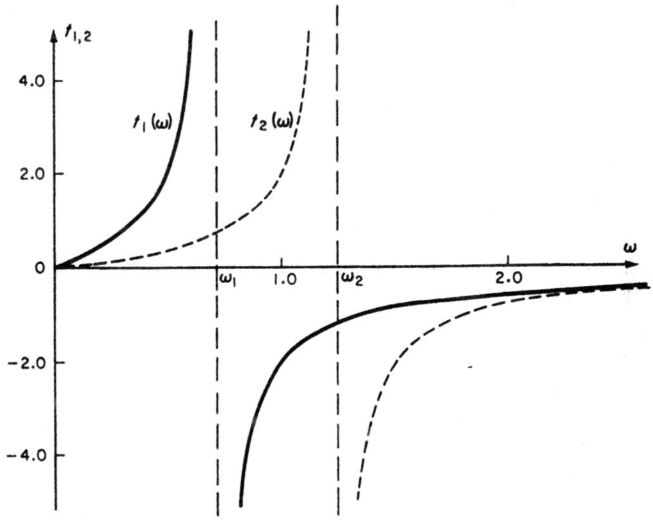

FIG. 8.1. Positive definite damping matrix.

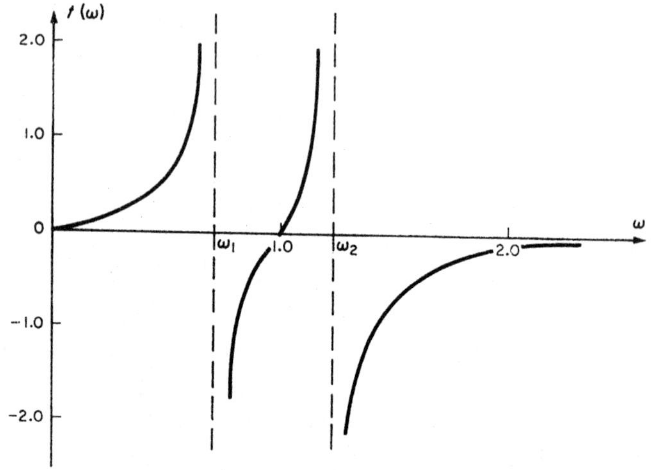

FIG. 8.2. Non-negative definite damping matrix.

8.4 Determination of the Natural Frequencies

Until recent years the practical estimation of the natural frequencies of lightly damped systems has usually been performed by observation of the resonant peaks—as described in § 8.1. Another method is now available which appears to originate with Asher [1], and which is much more satisfactory, from a theoretical point of view at least. In view of the discussion of § 8.3 it is convenient to assume now that $\omega B + G$ is positive definite for all $\omega \geqq 0$.

The method is based on the following theorem; we will discuss the application when the proof is complete.

THEOREM 8.4 *If the matrix $\omega B + G$ is positive definite and if $R(\omega)$ is the real part of the receptance matrix* (p. 147), *then* $|R(\omega)| = 0$ *if and only if* $\omega = \omega_s$.

Proof From (8.2.11) we have, for $\omega \neq \omega_s$,

$$R = VP(I + T^2)^{-1/2}\, V'. \tag{8.4.1}$$

Now as $\omega \rightarrow \omega_s$, $t_r(\omega) \rightarrow \pm\infty$ for some r and if ω_s has multiplicity α, then there are α proper numbers $t_r(\omega)$ which have singularities at $\omega = \omega_s$. (Note that this depends on the assumption that $\omega B + G$ is positive definite.) Theorem 8.2 implies that for $\omega \geqq 0$ there are α independent associated vectors to complete the matrix V. Thus, provided $\omega B + G$ is positive definite, (8.4.1) may be said to define R for *all* non-negative values of ω, and for all such ω, $|V| \neq 0$. From eqn. (8.2.13) we deduce that $|V|$ is also finite. Taking determinants in (8.4.1) we now observe that $|R| = 0$ if and only if there exists a proper number, $t_r(\omega) = \pm\infty$, that is, if and only if $\omega = \omega_s$.

This result may be applied in the following way. In (8.2.1) let $\boldsymbol{f}_0 = \boldsymbol{k}$, a *real* vector. That is, $\boldsymbol{k}$ represents forces applied at the frequency ω which are all exactly in phase with each other. Then, if $\boldsymbol{x}$ is the real part of the response vector, we find from (8.2.10) that

$$\boldsymbol{x}(\omega) = R(\omega)\,\boldsymbol{k}, \tag{8.4.2}$$

and the vector $\boldsymbol{x}$ is easily measured for a range of values of ω.

We now repeat the process n times at a fixed ω, but using linearly independent real force vectors $\boldsymbol{k}_1, \boldsymbol{k}_2, \ldots, \boldsymbol{k}_n$, and we measure the in-phase portion of the response vectors $\boldsymbol{x}_1, \boldsymbol{x}_2, \ldots, \boldsymbol{x}_n$. If

$$K = [\boldsymbol{k}_1, \boldsymbol{k}_2, \ldots, \boldsymbol{k}_n], \qquad \text{and} \qquad X = [\boldsymbol{x}_1, \boldsymbol{x}_2, \ldots, \boldsymbol{x}_n],$$

we then have

$$X(\omega) = R(\omega)\,K, \tag{8.4.3}$$

and since K has rank n, the above theorem implies that $|X| = 0$ if and only if $|R| = 0$.

Thus, by measuring the matrix X and evaluating $|X|$ as a function of ω, the natural frequencies are then found to be the zeros of $|X(\omega)|$.

In practice it may be useful to compute not $|X|$ but the smaller eigenvalues of X, and consider the behavior of these as functions of ω, making use of the fact that $|X| = 0$ if and only if X has a null eigenvalue. Such computations would be more laborious than simply computing the determinant but would probably clarify the problems where natural frequencies cluster together.

8.5 Determination of the Natural Modes

In this section we will discuss a method for determining the natural modes of vibration for the undamped system; but, as one might expect, the method presupposes a knowledge of the related natural frequencies. It will be assumed that these have been found by the method expounded above, or by any other suitable method.

We first prove the following important theorem, in which it is to be understood that ω_s is a typical natural frequency of the undamped system.

THEOREM 8.5 *The response* $\boldsymbol{q}e^{i\omega_s t}$ *to a disturbing force* $\boldsymbol{F}e^{i\omega_s t}$ *is such that* $\boldsymbol{q}$ *is a principal vector (unnormalized) if and only if* (i) *the elements of* $\boldsymbol{F}$ *are in the same phase, and* (ii) *the elements of* $\boldsymbol{q}$ *and* $\boldsymbol{F}$ *are* 90° *out of phase.*

Proof We first prove that these conditions are necessary. Let the response be proportional to the principal vector $\boldsymbol{q}_s$. Now from Theorem 8.2 we know that there is a proper vector $\boldsymbol{v}_r(\omega_s)$, say, which coincides with $\boldsymbol{q}_s$. Thus, for any ω we write

$$\boldsymbol{q} = \boldsymbol{v}_r e^{i\theta} \tag{8.5.1}$$

in which θ is real, and in which $\boldsymbol{q}$, $\boldsymbol{v}_r$ are all functions of ω, but are such that $\boldsymbol{v}_r(\omega_s) = \boldsymbol{q}_s$. From (8.2.9) we obtain, for $\omega \neq \omega_s$,

$$\boldsymbol{F} = (V')^{-1}(I + T^2)^{-1/2} P(I + iT) V^{-1}\boldsymbol{v}_r e^{i\theta},$$

and using (8.2.13),

$$\boldsymbol{F} = H(I + T^2)^{-1/2} P(I + iT) \boldsymbol{e}_r e^{i\theta},$$

where $\boldsymbol{e}_r$ has a 1 in the rth position and zeros elsewhere. Using the definitions of the diagonal matrices P and T we get

$$\boldsymbol{F} = (\operatorname{sgn} t_r) \left\{\frac{1 + it_r}{(1 + t_r^2)^{1/2}}\right\} \boldsymbol{h}_r e^{i\theta}.$$

Now we have seen that $t_r \to \pm\infty$ as $\omega \to \omega_s$, so that in the limit

$$\boldsymbol{F} = i\boldsymbol{h}_r e^{i\theta}, \tag{8.5.2}$$

where $\boldsymbol{h}_r$ ($\equiv \boldsymbol{h}_r(\omega_s)$) is real. Comparing (8.5.1) and (8.5.2) the propositions (i) and (ii) follow immediately.

Conservely, given (i) and (ii) we write $\boldsymbol{F} = i\boldsymbol{F}_1$ where $\boldsymbol{F}_1$ is real and we assume $\boldsymbol{q}$ to be real. From (8.2.1) we find that

$$\{(C - A\omega_s^2) + i(\omega_s B + G)\} \boldsymbol{q} = i\boldsymbol{F}_1.$$

Comparing real and imaginary parts we obtain

$$(C - A\omega_s^2) \boldsymbol{q} = \boldsymbol{0},$$
$$(\omega_s B + G) \boldsymbol{q} = \boldsymbol{F}_1. \tag{8.5.3}$$

The first of these equations implies that $\boldsymbol{q}$ is a principal vector, and completes the proof.

The following application of Theorem 8.5 originates with Traill-Nash [1]. We apply forces defined by the real parts of the vectors $\boldsymbol{l}_r e^{i\omega_s t}$, $r = 1, 2, \ldots, n$, where the $\boldsymbol{l}_r$ form a complete set of *real* vectors. Define

$$L = [\boldsymbol{l}_1, \boldsymbol{l}_2, \ldots, \boldsymbol{l}_n],$$

and we then have $|L| \neq 0$. We represent the responses to the above force distributions by

$$\boldsymbol{q}\, e^{i\omega_s t} = (\boldsymbol{x}_r + i\boldsymbol{y}_r)\, e^{i\omega_s t},$$

where the vectors $\boldsymbol{x}_r$ and $\boldsymbol{y}_r$ are real. We suppose that the vectors $\boldsymbol{x}_r$ are measured, and that L is given. Let

$$X = [\boldsymbol{x}_1, \boldsymbol{x}_2, \ldots, \boldsymbol{x}_n].$$

Let $\boldsymbol{f}$ be the force vector which gives rise to a response in the sth undamped normal mode, and define the vector $\boldsymbol{a}$ by

$$\boldsymbol{f} = L\boldsymbol{a}. \tag{8.5.4}$$

By Theorem 8.5, the real part of the response vector due to (the real part of) the force vector $\boldsymbol{f}e^{i\omega_s t}$ is zero. But the response vector arising from the application of $\boldsymbol{f}$ is the sum of those due to $\boldsymbol{l}_r a_r$, $r = 1, 2, \ldots, n$, and the real part of this response vanishes if and only if

$$\sum_{r=1}^{n} \boldsymbol{x}_r a_r = X\boldsymbol{a} = \boldsymbol{0}. \tag{8.5.5}$$

Thus, having measured X, $\boldsymbol{a}$ is found to belong to the null-space of X, and $\boldsymbol{f}$ may then be calculated from (8.5.4). This force distribution may then be applied to the system when the measured response should be exactly out of phase with $\boldsymbol{f}$, and be in a principal vector of ω_s.

It can also be shown without difficulty that the degeneracy of X is equal to the multiplicity of the proper number t_r, and hence (in general) of the natural frequency ω_s. That is, if there are α linearly independent principal vectors associated with ω_s, then there exist α linearly independent solutions of (8.5.5).

Obvious difficulties will arise in practice from the fact that the elements of X can only be measured to a limited accuracy, and $|X|$ will therefore not be exactly zero. In this event it may be sufficient to approximate the solutions $\boldsymbol{a}$ by the eigenvectors of those eigenvalues of X with the smallest modulus.

Appendix to Chapter 8

Theorem 8.6 *If A and C are real symmetric matrices and A is non-negative definite, then all the finite latent roots are real.*

Proof In order that the latent root problem should be non-trivial we take it that $|A\lambda + C| \not\equiv 0$, otherwise every scalar λ is a latent root. Observe first of all that, if $\boldsymbol{x}$ is an arbitrary vector and $\bar{\boldsymbol{x}}'A\boldsymbol{x} = 0$, then $A\boldsymbol{x} = \boldsymbol{0}$. In order to verify this we observe that there exists a real symmetric matrix $A^{1/2}$ (defined by eqn. (1.8.8)) such that $(A^{1/2})^2 = A$. Then $\bar{\boldsymbol{x}}'A\boldsymbol{x} = 0$ implies that $\bar{\boldsymbol{y}}'\boldsymbol{y} = 0$ where $\boldsymbol{y} = A^{1/2}\boldsymbol{x}$. This clearly implies that

$$\boldsymbol{y} = A^{1/2}\boldsymbol{x} = \boldsymbol{0},$$

whence $A\boldsymbol{x} = \boldsymbol{0}$.

Suppose now that there exist complex λ and $\boldsymbol{x}$ such that

$$(A\lambda + C)\,\boldsymbol{x} = \boldsymbol{0}, \tag{A.1}$$

where $\lambda = \mu + i\omega$, and $\omega \neq 0$. Premultiplying by $\bar{\boldsymbol{x}}'$ and noting that $\bar{\boldsymbol{x}}'C\boldsymbol{x}$ is real and $\bar{\boldsymbol{x}}'A\boldsymbol{x} \geqq 0$ (eqn. (1.5.7)), we deduce that $\bar{\boldsymbol{x}}'C\boldsymbol{x} = 0$ and hence $\bar{\boldsymbol{x}}'A\boldsymbol{x} = 0$. We have seen that this implies $A\boldsymbol{x} = \boldsymbol{0}$ and (A.1) then gives $C\boldsymbol{x} = \boldsymbol{0}$. But if there exists such an $\boldsymbol{x}$, then $|A\lambda + C| \equiv 0$ and we have excluded this possibility. Thus the assumption that λ and $\boldsymbol{x}$ are complex leads to a contradiction.

Let us now exhibit a pencil of the form (8.2.6) which is not simple. We take

$$A = \begin{bmatrix} 1 & 0 \\ 0 & 1 \end{bmatrix}, \qquad C = \begin{bmatrix} 2 & 0 \\ 0 & 1 \end{bmatrix},$$

$$B = \sqrt{\tfrac{2}{3}}\begin{bmatrix} 1 & -1 \\ -1 & 1 \end{bmatrix}, \qquad G = \begin{bmatrix} 1 & -1 \\ -1 & 1 \end{bmatrix}.$$

Note that G has eigenvalues 0, 2 and that B and G are non-negative definite. At $\omega^2 = 3/2$ we have

$$C - A\omega^2 = \begin{bmatrix} \frac{1}{2} & 0 \\ 0 & -\frac{1}{2} \end{bmatrix}, \qquad \omega B + G = \begin{bmatrix} 2 & -2 \\ -2 & 2 \end{bmatrix},$$

and the pencil (8.2.6) is proportional to

$$\begin{bmatrix} t - 4 & 4 \\ 4 & -t - 4 \end{bmatrix}$$

whose determinant is $-t^2$. Thus, the pencil has a double latent root, $t = 0$, and the nullspace of the matrix

$$\begin{bmatrix} -4 & 4 \\ 4 & -4 \end{bmatrix}$$

has dimension one. The pencil is therefore defective at this value of ω.

CHAPTER 9

FURTHER RESULTS FOR SYSTEMS WITH DAMPING

9.1 PRELIMINARIES

In this chapter we will be concerned with the nature of solutions of the matrix differential equation:

$$A\ddot{\boldsymbol{p}} + \varepsilon B\dot{\boldsymbol{p}} + C\boldsymbol{p} = \boldsymbol{0}, \tag{9.1.1}$$

where ε is a real non-negative number and:

(i) A is a real symmetric, positive definite matrix of order n,
(ii) B is a real matrix of order n,
(iii) C is a real symmetric, non-negative definite matrix of order n.

We already have some information in special cases. The reader's attention is drawn to the results of § 7.5, § 7.6, and § 7.7. Our point of view for the greater part of this chapter will be that of perturbation theory, ε being the perturbation parameter. This implies a prior knowledge of the system when $\varepsilon = 0$, as discussed in § 7.3. It is therefore assumed that the reader is familiar with the results and terminology of that section and that the matrices Q and W of eqns. (7.3.5) are known. That is, we have a real non-singular matrix Q and (in a modified notation) a real diagonal matrix $W = \text{diag}\,\{\omega_{10}, \omega_{20}, \ldots, \omega_{n0}\}$ such that

$$Q'AQ = I, \qquad Q'CQ = W^2, \tag{9.1.2}$$

and

$$0 \leqq \omega_{10} \leqq \omega_{20} \leqq \cdots \leqq \omega_{n0}. \tag{9.1.3}$$

The latter numbers are the *undamped natural frequencies*.

On transforming to principal coordinates $q_1, q_2, \ldots, q_n$ defined by

$$\boldsymbol{q} = Q^{-1}\boldsymbol{p}$$

and using eqns. (9.1.2), equation (9.1.1) reduces to

$$\ddot{\boldsymbol{q}} + \varepsilon\beta\dot{\boldsymbol{q}} + W^2\boldsymbol{q} = \boldsymbol{0}, \tag{9.1.4}$$

where

$$\beta = Q'BQ. \tag{9.1.5}$$

Now we have seen in § 6.2 that, if $D(\lambda) = A\lambda^2 + \varepsilon B\lambda + C$ is a simple λ-matrix, then the solution space of equation (9.1.1) is spanned by all vectors of the form $e^{\lambda t}\boldsymbol{x}$, where λ is a latent root of $D(\lambda)$ with right latent vector $\boldsymbol{x}$. We immediately deduce that, on the same assumption, the solution space of (9.1.4) is spanned by all vectors of the form $e^{\lambda t}\boldsymbol{q}$, where λ, $\boldsymbol{q}$ are a set of (right) latent values of

$$(I\lambda^2 + \varepsilon\beta\lambda + W^2)\,\boldsymbol{q} = \boldsymbol{0}. \tag{9.1.6}$$

When $\varepsilon = 0$ the latent roots are $\pm\, i\omega_{r0}$, $r = 1, 2, \ldots, n$. If we write (for any ε) $\lambda_r = \mu_r + i\omega_r$, where μ_r, ω_r are real, then we refer to μ_r, ω_r as the *rate of decay* and the *frequency* of the rth mode of free vibration respectively.

We now consider a matrix whose eigenvalues coincide with the latent roots of (9.1.6), namely the $2n \times 2n$ matrix

$$L = \begin{bmatrix} -\varepsilon\beta & iW \\ iW & 0 \end{bmatrix}, \tag{9.1.7}$$

which is suggested by (4.2.6). It is easily verified that

$$(L - \lambda I)\begin{bmatrix} \boldsymbol{q} \\ \boldsymbol{q}_1 \end{bmatrix} = \boldsymbol{0}$$

implies (9.1.6) together with

$$\lambda\boldsymbol{q}_1 = iW\boldsymbol{q}.$$

We also recall that any matrix which is similar to L will have the same eigenvalues as L and hence have eigenvalues coincident with the λ_r (cf. § 1.8). In particular, if we consider the orthogonal matrix

$$T = \frac{1}{\sqrt{2}}\begin{bmatrix} I & I \\ -I & I \end{bmatrix} \tag{9.1.8}$$

and let $L_1 = TLT^{-1}$, then it is found that

$$L_1 = \begin{bmatrix} iW - \frac{1}{2}\varepsilon\beta & \frac{1}{2}\varepsilon\beta \\ \frac{1}{2}\varepsilon\beta & -iW - \frac{1}{2}\varepsilon\beta \end{bmatrix} \tag{9.1.9}$$

has its eigenvalues coincident with the latent roots of (9.1.6).

9.2 Global Bounds for the Latent Roots when B is Symmetric

In this section we obtain bounds for the latent roots when B is real and symmetric. We will see that very simple geometrical bounds can be described in the complex λ-plane.

Let λ be a latent root of (9.1.6) with latent vector $\boldsymbol{q}$. We may suppose without loss of generality that $\bar{\boldsymbol{q}}'\boldsymbol{q} = 1$. Then premultiplying (9.1.6) by $\bar{\boldsymbol{q}}'$ we obtain

$$\lambda^2 + \varepsilon b\lambda + w^2 = 0, \qquad (9.2.1)$$

where

$$b = \bar{\boldsymbol{q}}'\beta\boldsymbol{q}, \qquad w^2 = \bar{\boldsymbol{q}}'W^2\boldsymbol{q}. \qquad (9.2.2)$$

Now the assumption that B is symmetric and (9.1.5) together imply that β is symmetric and hence b is real. Clearly, we also have $w^2 \geqq 0$. Writing $\lambda = \mu + i\omega$ in (9.2.1) and equating real and imaginary parts of the left-hand side to zero we obtain

$$\mu = -\tfrac{1}{2}\varepsilon b \qquad (9.2.3)$$

and

$$w^2 = \omega^2 - \mu^2 - \varepsilon b\mu = \omega^2 + \mu^2 = |\lambda|^2, \qquad (9.2.4)$$

provided $\omega \neq 0$. Now the variational characterization of eigenvalues (inequality (1.7.3)) implies that

$$\beta_1 \leqq b \leqq \beta_n \qquad \text{and} \qquad \omega_{10}^2 \leqq w^2 \leqq \omega_{n0}^2$$

where $\beta_1, \beta_2, \dots, \beta_n$ are the eigenvalues of β in non-decreasing order. Thus, (9.2.3) and (9.2.4) imply that, for any complex latent root,

$$-\tfrac{1}{2}\varepsilon\beta_n \leqq \mu \leqq -\tfrac{1}{2}\varepsilon\beta_1 \qquad (9.2.5)$$

and

$$\omega_{10} \leqq |\lambda| \leqq \omega_{n0}. \qquad (9.2.6)$$

The latter inequality implies that, whatever the value of ε, all the complex latent roots lie in an annulus of the λ-plane. The former inequality provides a vertical strip of the λ-plane which must also contain all the complex latent roots. A case in which $0 < \beta_1 < \beta_n$ and $0 < \omega_{10} < \omega_{n0}$ is sketched in Fig. 9.1. All the complex latent roots must lie in the hatched area. In this case it is easily seen that, if

$$\varepsilon > 2\omega_{n0}/\beta_1 \qquad (9.2.7)$$

then all the latent roots are real, and, if

$$\varepsilon < 2\omega_{10}/\beta_n, \qquad (9.2.8)$$

then all the latent roots are complex.

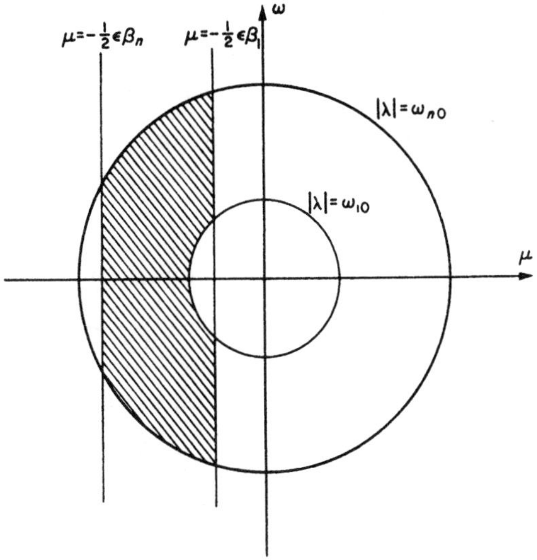

FIG. 9.1. Bounds for the complex latent roots.

For the real latent roots, if any, the following inequalities may be deduced:

(a) For negative roots

$$-\tfrac{1}{2}\varepsilon\beta_n - \sqrt{(\tfrac{1}{2}\varepsilon\beta_n)^2 - \omega_{10}^2} \leqq \lambda \leqq -\tfrac{1}{2}\varepsilon\beta_n + \sqrt{(\tfrac{1}{2}\varepsilon\beta_n)^2 - \omega_{10}^2}\,. \tag{9.2.9}$$

(b) For positive roots

$$-\tfrac{1}{2}\varepsilon\beta_1 - \sqrt{(\tfrac{1}{2}\varepsilon\beta_1)^2 - \omega_{10}^2} \leqq \lambda \leqq -\tfrac{1}{2}\varepsilon\beta_1 + \sqrt{(\tfrac{1}{2}\varepsilon\beta_1)^2 - \omega_{10}^2}\,.$$

Note that when B is non-negative, or positive definite, all the real roots satisfy (a), and in particular, $-\varepsilon\beta_n \leqq \lambda \leqq 0$. When B is negative, or non-positive definite, all the real roots satisfy (b). It should also be noted that if $(\tfrac{1}{2}\varepsilon\beta_n)^2 < \omega_{10}^2$ then there are no negative real roots, and if $(\tfrac{1}{2}\varepsilon\beta_1)^2 < \omega_{10}^2$ then there are no positive real roots.

The eigenvalues $\beta_1, \beta_2, \ldots, \beta_n$ may be interpreted in a second illuminating way. We have defined these numbers as the eigenvalues of $\beta = Q'BQ$. Thus, $|Q'BQ - \beta_i I| = 0$. But this implies that $|B - \beta_i(Q')^{-1}Q^{-1}| = 0$, and then the first of equations (9.1.2) implies that $(Q')^{-1}Q^{-1} = A$, so that $|B - \beta_i A| = 0$, and the *eigenvalues of* β *therefore coincide with the latent roots of the pencil* $B - \lambda A$. We then deduce from (2.8.4) that, for any real $\boldsymbol{q} \neq \boldsymbol{0}$,

$$\beta_1 \leqq \frac{q'Bq}{q'Aq} \leqq \beta_n.$$

Schur's inequalities for eigenvalues (Mirsky [1]) (which include bounds due to Hirsch) may also be applied to L or L_1. However, they give no improvement on the results of this and the following section.

9.3 The Use of Theorems on Bounds for Eigenvalues

We immediately obtain bounds for the latent roots by applying Theorems 1.6, 1.7, and 1.8 to the matrix L of (9.1.7) or to any matrix obtained from L by a similarity transformation; matrix L_1 of (9.1.9) for example. We first apply Geršgorin's theorem (1.6) to L_1. The radii of the Geršgorin circles are obtained from (1.10.1) in the form $\varepsilon\sigma_r$ where

$$\sigma_r = \tfrac{1}{2}|\beta_{rr}| + \sum_{\substack{s=1\\ s\neq r}}^{n} |\beta_{rs}|, \tag{9.3.1}$$

for $r = 1, 2, \ldots, n$. The centres of the circles are at the points $\pm i\omega_{r0} - \frac{1}{2}\varepsilon\beta_{rr}$. When the circular bounds are all sketched in the λ-plane there will be symmetry with respect to the real axis, and we may therefore focus attention on the upper half-plane and the real axis only. The latent roots in this half-plane are contained in the union of the circular regions defined by

$$|\lambda - (i\omega_{r0} - \tfrac{1}{2}\varepsilon\beta_{rr})| \leqq \varepsilon\sigma_r, \qquad r = 1, 2, \ldots, n. \tag{9.3.2}$$

If B is not symmetric, then application of the theorem to the columns of L_1, rather than the rows as used above, will provide a second set of circles having the same centres, but generally different radii.

A typical circle is sketched in Fig. 9.2, and, if such a circle has no point in common with the other circles, then it contains one and only one latent root of (9.1.6). Thus, writing $\lambda_r = \mu_r + i\omega_r$ and assuming $\beta_{rr} > 0$ we would have

$$\varepsilon(\beta_{rr} + \sum_{\substack{s=1\\ s\neq r}}^{n} |\beta_{rs}|) \leqq \mu_r \leqq \varepsilon \sum_{\substack{s=1\\ s\neq r}}^{n} |\beta_{rs}|, \tag{9.3.3}$$

and

$$\omega_{r0} - \varepsilon\sigma_r \leqq \omega_r \leqq \omega_{r0} + \varepsilon\sigma_r. \tag{9.3.4}$$

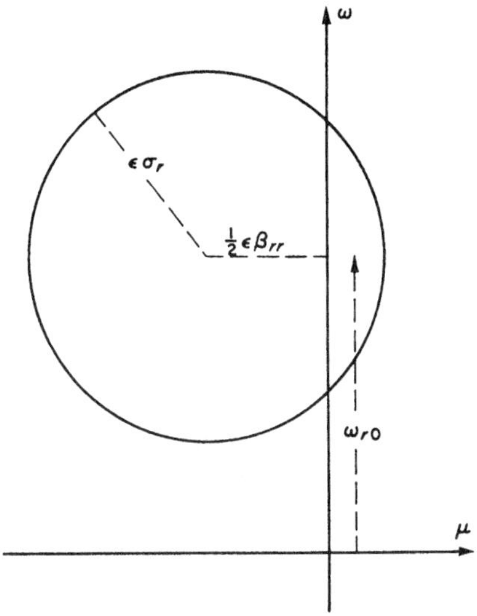

FIG. 9.2. A typical disc obtained by Geršgorin's theorem.

The main distinction between these and subsequent results and those of § 9.2 lies in the fact that we may now be able to give bounds for the individual latent roots, whereas the analysis of § 9.2 does not permit this.

Theorem 1.7 now suggests itself in cases where ε is small, for the dimensions of the ovals (1.10.3) are determined by products of the quantities $\varepsilon\sigma_r$, and are generally proportional to ε^2. This is not the case, however, for an oval produced when $\omega_{r0} = \omega_{s0}$ $(r \neq s)$. In these circumstances the bounds for the roots which coincide at $\varepsilon = 0$ are no better than those obtained above, though useful bounds can be obtained for other roots which are not repeated at $\varepsilon = 0$.

The ovals obtained by applying the theorem to rows r and s of L_1 (eqn. (9.1.9)), where $r, s, \leqq n$ are given by

$$|\lambda - (i\omega_{r0} - \tfrac{1}{2}\varepsilon\beta_{rr})|\,|\lambda - (i\omega_{s0} - \tfrac{1}{2}\varepsilon\beta_{ss})| \leqq \varepsilon^2\sigma_r\sigma_s. \quad (9.3.5)$$

If $\omega_{r0} \neq \omega_{s0}$, there are two distinct regions of the λ-plane in which this inequality is satisfied (see inequality (1.10.4) and Fig. 1.1) provided

$$2\varepsilon\sqrt{\sigma_r\sigma_s} < |(i\omega_{r0} - \tfrac{1}{2}\varepsilon\beta_{rr}) - (i\omega_{s0} - \tfrac{1}{2}\varepsilon\beta_{ss})|,$$

and this is certainly the case if

$$\varepsilon\sqrt{\sigma_r\sigma_s} \leqq \tfrac{1}{2}|\omega_{r0} - \omega_{s0}| = d_{rs}, \tag{9.3.6}$$

say. Furthermore, if circles with centres at $i\omega_{r0} - \frac{1}{2}\varepsilon\beta_{rr}$ and $i\omega_{s0} - \frac{1}{2}\varepsilon\beta_{ss}$ and having radius $\varepsilon^2 R_{rs}$, where

$$\varepsilon^2 R_{rs} = d_{rs} - (d_{rs}^2 - \varepsilon^2\sigma_r\sigma_s)^{1/2} = \frac{\varepsilon^2\sigma_r\sigma_s}{d_{rs} + (d_{rs}^2 - \varepsilon^2\sigma_r\sigma_s)^{1/2}}, \tag{9.3.7}$$

are drawn, then they each contain one of the above oval regions and they do not intersect. These circles have a radius which is $O(\varepsilon^2)$ as $\varepsilon \to 0$.

We may simplify the analysis by choosing for our bounds circles whose radius exceeds $\varepsilon^2 R_{rs}$. For example, we may choose circles of radius

$$\varepsilon^2 R_{rs}^* = \varepsilon^2 \frac{\sigma_r\sigma_s}{d_{rs}}. \tag{9.3.8}$$

Now there are $n - 1$ such circles with centre $i\omega_{r0} - \frac{1}{2}\varepsilon\beta_{rr}$, and they are all contained in the circle of radius

$$\varepsilon^2\sigma_r \max_s\left(\frac{\sigma_s}{d_{rs}}\right), \qquad s \neq r,$$

which does not exceed

$$\varepsilon^2 \frac{\sigma_r\sigma_m}{\frac{1}{2}\min_s|\omega_{r0} - \omega_{s0}|}, \qquad s \neq r,$$

where $\sigma_m = \max(\sigma_r)$. There are further points to be clarified concerning the ovals obtained for $r \leqq n$ and $s > n$, but we omit the details and claim the following:

THEOREM 9.1 *If all the undamped natural frequencies of* (9.1.6) *are distinct and*

$$\varepsilon\sqrt{\sigma_r\sigma_s} < \tfrac{1}{2}|\omega_{r0} - \omega_{s0}|, \qquad r, s = 1, 2, \ldots, n, \quad r \neq s,$$

and

$$\varepsilon\sigma_r < \omega_{r0},$$

then all the latent roots of (9.1.6) *lie in the union of the circular regions*

$$|\lambda - (i\omega_{r0} - \tfrac{1}{2}\varepsilon\beta_{rr})| < \varepsilon^2 R_r \tag{9.3.9}$$

where

$$R_r = \sigma_r\sigma_m/A_r$$

and

$$\sigma_m = \max_r(\sigma_r), \qquad A_r = \tfrac{1}{2}\min\{\min_{\substack{s \\ s \neq r}} |\omega_{r0} - \omega_{s0}|, 2\omega_{10}\}.$$

Cruder, but more easily computed, bounds are obtained if we take circles of the same radius R, such that $R > \max_r (R_r)$. In particular, we may take

$$R = \sigma_m^2/\min_r (A_r) = \frac{2\sigma_m^2}{\min\{\min_{\substack{r\,s \\ r \neq s}} |\omega_{r0} - \omega_{s0}|, 2\omega_{10}\}}.$$

We may immediately deduce that, under the conditions of the theorem,

$$\lambda_r = i\omega_{r0} - \tfrac{1}{2}\varepsilon\beta_{rr} + O(\varepsilon^2)\dagger \tag{9.3.10}$$

as $\varepsilon \to 0$, a result which we will confirm by classical perturbation theory in § 9.5.

Finally, we observe how Theorem 1.8 may be applied. We first write $L = M + N$ where

$$M = \begin{bmatrix} 0 & iW \\ iW & 0 \end{bmatrix}, \qquad N = \begin{bmatrix} -\varepsilon\beta & 0 \\ 0 & 0 \end{bmatrix},$$

and assume that B is real and symmetric, in which case β and N are also real and symmetric, and are therefore unitarily similar to diagonal matrices (in fact, the transforming matrix may be real orthogonal and hence unitary). The real orthogonal matrix T of (9.1.8) is such that TMT' is diagonal, so that M is also unitarily similar to a diagonal matrix. We may therefore apply Theorem 1.8 to this decomposition of L, observing that, in our previous notation, the eigenvalues of N are

$$-\varepsilon\beta_n \leqq \cdots \leqq -\varepsilon\beta_2 \leqq -\varepsilon\beta_1$$

together with n zeros, and the eigenvalues of M are $\pm i\omega_{r0}$, $r = 1, 2, \ldots, n$. In accordance with our discussion of Theorem 1.8, we choose for the parameter α the arithmetic mean of the extreme eigenvalues of N. Thus, if the extreme eigenvalues are $\varepsilon\gamma$, $\varepsilon\delta$, then

$$\gamma = \min(0, -\beta_n), \qquad \delta = \max(0, -\beta_1)$$

† Using a technique described by Wilkinson [3], this result can also be obtained directly from Geršgorin's theorem.

and we choose

$$\alpha = \tfrac{1}{2}\varepsilon(\gamma + \delta)$$

so that

$$\varrho = \tfrac{1}{2}\varepsilon(\delta - \gamma).$$

The theorem then states that all the eigenvalues of L (latent roots of (9.1.6)) lie in the union of the circular regions:

$$|\lambda - \{i\omega_{r0} + \tfrac{1}{2}\varepsilon(\gamma + \delta)\}| \leqq \tfrac{1}{2}\varepsilon(\delta - \gamma), \qquad r = 1, 2, \ldots, n. \quad (9.3.11)$$

Let us illustrate the theorem when B, and hence β, is non-negative or positive definite. In this case we have $\beta_1 \geqq 0$ and

$$\gamma = -\beta_n, \qquad \delta = 0.$$

Thus, the circular regions (9.3.10) become

$$|\lambda - (i\omega_{r0} - \tfrac{1}{2}\varepsilon\beta_n)| \leqq \tfrac{1}{2}\varepsilon\beta_n, \qquad (9.3.12)$$

and the circular boundaries are tangent to the imaginary axis in the λ-plane. In practice, several of the bounds discussed could be used simultaneously in order to improve estimates of the latent roots. In particular, if the lower bound of (9.2.5) is combined with the above, the set of circles is immediately reduced to a set of *semicircles*, and we obtain a considerable improvement.

It should be observed that (9.2.5), (9.3.11) and (9.3.12) may yield useful bounds without explicit calculation of the eigenvalues of β. Let us suppose that β_n, as well as being the greatest eigenvalue of β, is also the eigenvalue of greatest modulus. There exist certain measures of magnitude of a matrix known as multiplicative matrix norms (Ostrowski [1], [3], Householder [1], Stone [1]), and it can be proved that, if A is a matrix with eigenvalues $\mu_1, \mu_2, \ldots, \mu_n$, $N(A)$ is such a norm of A, and

$$\mu_M = \max_r |\mu_r|, \qquad r = 1, 2, \ldots, n,$$

then

$$N(A) \geqq \mu_M.$$

Furthermore, if A, B are square matrices of the same order, then

$$N(AB) \leqq N(A)\, N(B).$$

Thus in the bounds we have obtained we could replace β_n by $N(\beta)$ where N is an easily computed norm. However, we have $\beta = Q'BQ$, and the explicit computation of β may also be avoided if we observe that

$$\beta_n \leqq N(\beta) \leqq N(Q')\, N(B)\, N(Q).$$

The most convenient norms are likely to be among the following. In each case we have $i, j = 1, 2, \ldots, n$.

$$N_1(A) = \max_i \sum_j |a_{ij}|,$$

$$N_2(A) = \max_i \sum_j |a_{ij}|,$$

$$N_3(A) = n \max_{i,j} |a_{ij}|,$$

$$N_4(A) = \{\sum_{i,j} |a_{ij}|^2\}^{1/2},$$

$$N_5(A) = \sum_{i,j} |a_{ij}|.$$

Let us illustrate some of the results of this and the preceding sections with a numerical example due to Falk [1]. In (9.1.1) we take

$$A = \begin{bmatrix} 1 & 0 & 0 \\ 0 & 2 & 0 \\ 0 & 0 & 5 \end{bmatrix}, \quad B = \begin{bmatrix} 0 & 0 & 0 \\ 0 & 3 & -1 \\ 0 & -1 & 6 \end{bmatrix},$$

$$C = \begin{bmatrix} 2 & -1 & 0 \\ -1 & 3 & 0 \\ 0 & 0 & 10 \end{bmatrix}.$$

Investigation of the undamped system gives

$$W^2 = \text{diag}\,\{\omega_{10}^2, \omega_{20}^2, \omega_{30}^2\} = \begin{bmatrix} 1 & 0 & 0 \\ 0 & 2 & 0 \\ 0 & 0 & 2.5 \end{bmatrix}$$

and

$$Q = \begin{bmatrix} 1 & 0 & -\sqrt{6}/3 \\ 1 & 0 & \sqrt{6}/6 \\ 0 & \sqrt{5}/5 & 0 \end{bmatrix}$$

We then compute

$$\beta = Q'BQ = \begin{bmatrix} 3 & -\sqrt{3}/3 & \sqrt{6}/2 \\ -\sqrt{3}/3 & 2 & -\sqrt{2}/6 \\ \sqrt{6}/2 & -\sqrt{2}/6 & 1/2 \end{bmatrix}$$

The eigenvalues of this matrix coincide with the latent roots of $A\lambda - B$ and are found to be $\beta_1 = 0$, $\beta_2 = 1$, $\beta_3 = 1.7$. We deduce that β, and hence B, is non-negative definite.

In Fig. 9.3 we sketch (with a continuous line) the bounds obtained from (9.2.5) and (9.2.6) for $\varepsilon = 1$ and $\varepsilon = 0.2$. The bounds obtained by application of (9.3.12) are indicated with broken lines. The latter bounds give us very little further information in the

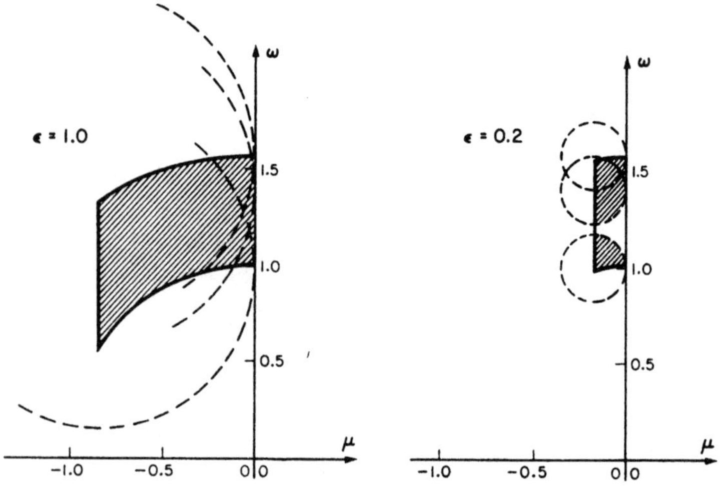

Fig. 9.3. Bounds for latent roots in the example with $\varepsilon = 1.0$, and $\varepsilon = 0.2$.

case $\varepsilon = 1$, but make a significant contribution when $\varepsilon = 0.2$. Note that, in the latter case, there is one root in the lower hatched area and there are two in the upper. In either case, the Geršgorin bounds (9.3.2) are too wide to be of any value, and the conditions imposed in Theorem 9.1 are not satisfied. However, these can be expected to yield the most useful results with smaller values of ε.

9.4 Preliminary Remarks on Perturbation Theory

We continue to discuss the problem formulated in § 9.1, and proceed more directly with analysis of the latent roots and vectors of (9.1.6) viewed as functions of ε in some neighborhood of $\varepsilon = 0$. We already have the important result (9.3.10) in this direction, though this is restricted to unrepeated, non-zero, undamped natural frequencies ω_{r0}. This result may also be anticipated by an application of Theorem 4.11, concerning the Rayleigh quotient, and provides an interesting application of that theorem.

Suppose that the latent root $\lambda_r(\varepsilon)$ of (9.1.6) is such that $\lambda_r(0) = i\omega_{r0}$, and the corresponding vector $\boldsymbol{x}_r(\varepsilon)$ is then such that $\boldsymbol{x}_r(0) = \boldsymbol{e}_r$, the vector with a 1 in the rth position and zeros elsewhere. Now it is known that $\lambda_r(\varepsilon)$, $\boldsymbol{x}_r(\varepsilon)$ may be assumed to be continuous in ε (see below) in some neighborhood of $\varepsilon = 0$, so we take $\boldsymbol{e}_r$ and $i\omega_{r0}$ as first approximations to $\boldsymbol{x}_r(\varepsilon)$, $\lambda_r(\varepsilon)$ for sufficiently small ε. Theorem 4.11 then suggests that

$$\lambda^* = i\omega_{r0} - \frac{\boldsymbol{e}_r'(I\omega_{r0}^2 + i\varepsilon\beta\omega_{r0} + W^2)\,\boldsymbol{e}_r}{\boldsymbol{e}_r'(2i\omega_{r0}I + \varepsilon\beta)\,\boldsymbol{e}_r}$$

will give a better estimate of $\lambda(\varepsilon)$. It is easily seen that this reduces to

$$\begin{aligned}\lambda^* &= i\omega_{r0} - \frac{i\varepsilon\omega_{r0}\beta_{rr}}{2i\omega_{r0} + \varepsilon\beta_{rr}} \\ &= i\omega_{r0} - \tfrac{1}{2}\varepsilon\beta_{rr} + 0(\varepsilon^2)\end{aligned}$$

as $\varepsilon \to 0$, as eqn. (9.3.10) has lead us to expect.

The classical perturbation theory, which we will apply in § 9.4, proceeds on the assumption that there exists a Taylor series expansion for each root $\lambda_r(\varepsilon)$ about $\varepsilon = 0$, and similarly for the elements of each latent vector. We may well ask to what extent this is justified. Partial, but very helpful, answers are provided by the discussion of § 2.4 if we recall that latent roots are also the eigenvalues of matrix L of (9.1.7). We immediately infer that, if ω_{r0} is an unrepeated undamped natural frequency, then $\lambda_r(\varepsilon)$ is a regular function of ε in some neighborhood of $\varepsilon = 0$. Unfortunately L is not a Hermitian matrix and the results of Rellich are not applicable for those $\lambda_r(\varepsilon)$ corresponding to a repeated natural frequency ω_{r0}. However, we will see that when β is real and symmetric Theorem 2.5 and 2.7 may be applied so that all the $\lambda_r(\varepsilon)$, without exception, are at least twice differentiable at $\varepsilon = 0$.

In spite of the danger of being repetitious, it is instructive to employ Theorems 2.5–2.7 at this stage in order to evaluate the derivatives $\lambda_r^{(1)}(0)$ and $\lambda_r^{(2)}(0)$. We consider the matrix L as a function of ε, and note first of all that $L(0)$ is a matrix of simple structure, and then that

$$L^{(1)}(0) = \begin{bmatrix} -\beta & 0 \\ 0 & 0 \end{bmatrix}, \qquad L^{(2)} = L^{(3)} = \cdots = 0. \qquad (9.4.1)$$

The eigenvalues of $L(0)$ are

$$\left.\begin{aligned}\lambda_r(0) &= i\omega_{r0}\\ \lambda_{n+r}(0) &= -i\omega_{r0}\end{aligned}\right\}, \qquad r = 1, 2, \ldots, n, \tag{9.4.2}$$

and the eigenvectors, defined to satisfy the hypothesis of Theorem 2.5, are

$$\boldsymbol{x}_r(0) = \frac{1}{\sqrt{2}}\begin{bmatrix}\boldsymbol{e}_r\\ \boldsymbol{e}_r\end{bmatrix}, \qquad \boldsymbol{x}_{n+r}(0) = \frac{1}{\sqrt{2}}\begin{bmatrix}\boldsymbol{e}_r\\ -\boldsymbol{e}_r\end{bmatrix}. \tag{9.4.3}$$

Suppose first of all that $\lambda_r(0) \neq 0$ and is an unrepeated eigenvalue of $L(0)$. Then part (ii) of Theorem 2.5 gives

$$\lambda_r^{(1)}(0) = \tfrac{1}{2}\,[\boldsymbol{e}_r' \quad \boldsymbol{e}_r']\begin{bmatrix}-\beta & 0\\ 0 & 0\end{bmatrix}\begin{bmatrix}\boldsymbol{e}_r\\ \boldsymbol{e}_r\end{bmatrix} = -\tfrac{1}{2}\beta_{rr}. \tag{9.4.4}$$

It is easily verified that coefficients p_{jk} of (2.4.13) are the elements of the matrix

$$P = -\frac{1}{2}\begin{bmatrix}\beta & \beta\\ \beta & \beta\end{bmatrix}$$

and, when β is symmetric, eqn. (2.4.12) gives

$$\lambda_r^{(2)}(0) = \frac{1}{2}\,\frac{\beta_{rr}^2}{\lambda_r - \lambda_{n+r}} + \frac{1}{2}\sum_{\substack{k=1\\ k\neq r}}^{n} \beta_{rk}^2\left(\frac{1}{\lambda_r - \lambda_k} + \frac{1}{\lambda_r - \lambda_{n+k}}\right),$$

the λ's all being evaluated at $\varepsilon = 0$. Noting, from (9.4.2), that $\lambda_{n+k} = -\lambda_k$ it is found that

$$\begin{aligned}\lambda_r^{(2)}(0) &= \frac{\beta_{rr}^2}{4\lambda_r} + \lambda_r\sum_{\substack{k=1\\ k\neq r}}^{n}\frac{\beta_{rk}^2}{\lambda_r^2 - \lambda_k^2}\\ &= -\frac{i\beta_{rr}^2}{4\omega_{r0}} - i\omega_{r0}\sum_{\substack{k=1\\ k\neq r}}^{n}\frac{\beta_{rk}^2}{\omega_{r0}^2 - \omega_{k0}^2}\end{aligned} \tag{9.4.5}$$

If $\lambda_r(0) \neq 0$ but is a repeated root the theorems still apply. For example, if $\lambda_1(0) = \lambda_2(0) \neq \lambda_3(0)$, and $\lambda_1(0) \neq 0$, then the first derivatives, $\lambda_{1,2}^{(1)}(0)$ are given by the eigenvalues of

$$\tfrac{1}{2}\begin{bmatrix}\boldsymbol{e}_1' & \boldsymbol{e}_1'\\ \boldsymbol{e}_2' & \boldsymbol{e}_2'\end{bmatrix}\begin{bmatrix}-\beta & 0\\ 0 & 0\end{bmatrix}\begin{bmatrix}\boldsymbol{e}_1 & \boldsymbol{e}_2\\ \boldsymbol{e}_1 & \boldsymbol{e}_2\end{bmatrix} = -\tfrac{1}{2}\begin{bmatrix}\beta_{11} & \beta_{12}\\ \beta_{21} & \beta_{22}\end{bmatrix}.$$

If β is real and symmetric, then the hypothesis of Theorem 2.7 is satisfied, and the second derivatives $\lambda_{1,2}^{(2)}(0)$ exist and may be computed as indicated in that theorem.

Finally, suppose that $\omega_{10} = 0$, $\omega_{20} \neq 0$, in which case $\lambda_1(0) = \lambda_{n+1}(0) = 0$, and these are the only null eigenvalues of $L(0)$. Then $\lambda_1^{(1)}(0)$ and $\lambda_{n+1}^{(1)}(0)$ are the eigenvalues of

$$\tfrac{1}{2}\begin{bmatrix} e_1' & e_1' \\ e_2' & -e_1' \end{bmatrix} \begin{bmatrix} -\beta & 0 \\ 0 & 0 \end{bmatrix} \begin{bmatrix} e_1 & e_1 \\ e_1 & -e_1 \end{bmatrix} = -\tfrac{1}{2}\begin{bmatrix} \beta_{11} & \beta_{11} \\ \beta_{11} & \beta_{11} \end{bmatrix}.$$

Thus, we have

$$\lambda_1^{(1)}(0) = 0, \qquad \lambda_{n+1}^{(1)}(0) = -\beta_{11}. \tag{9.4.6}$$

Once more, Theorem 2.7 may be applied and the second derivatives $\lambda_1^{(2)}(0)$ and $\lambda_{n+1}^{(2)}(0)$ may be computed. It is clear that there is one latent root which is identically zero. The assignment of suffixes in (9.4.6) implies that λ_1 is this null root, and that $\lambda_{n+1}(\varepsilon)$ is, for sufficiently small ε, a real latent root. It is found that

$$\lambda_1^{(2)}(0) = \lambda_{n+1}^{(2)}(0) = 0. \tag{9.4.7}$$

We will reproduce all of the above results, and more besides, with our perturbation analysis. However, we have achieved independent confirmation of the analysis, as well as a convincing illustration of Theorems 2.5, 2.6, and 2.7.

9.5 The Classical Perturbation Technique for Light Damping

We now proceed on the assumption that the latent roots $\lambda_r(\varepsilon)$ and the elements of the latent vectors $\boldsymbol{q}_r(\varepsilon)$ of eqn. (9.1.6) are regular functions of λ at $\varepsilon = 0$, or, at least, that they should have as many derivatives at $\varepsilon = 0$ as are required by the current discussion. The extent to which these assumptions are rigorously justified is outlined in § 2.4 and § 9.4. For the Taylor series of $\lambda_r(\varepsilon)$ about $\varepsilon = 0$ we write

$$\lambda_r(\varepsilon) = \lambda_{0r} + \varepsilon\lambda_{1r} + \varepsilon^2\lambda_{2r} + \dots, \tag{9.5.1}$$

so that

$$\lambda_{sr} = \frac{1}{s!}\lambda_r^{(s)}(0), \qquad s = 1, 2, \dots, \tag{9.5.2}$$

and

$$\left.\begin{aligned} \lambda_{0r} &= i\omega_{r0} \\ \lambda_{0,\,n+r} &= -i\omega_{r0} \end{aligned}\right\}, \qquad r = 1, 2, \dots, n. \tag{9.5.3}$$

If $\Lambda = \operatorname{diag}\{\lambda_1, \lambda_2, \dots, \lambda_n\}$, and $\Lambda_s = \operatorname{diag}\{\lambda_{s1}, \dots, \lambda_{sn}\}$, $s = 1, 2, \dots,$ then

$$\Lambda = \Lambda_0 + \varepsilon\Lambda_1 + \varepsilon^2\Lambda_2 + \dots. \tag{9.5.4}$$

Suppose that $\lambda_r(\varepsilon)$ has an associated latent vector $\boldsymbol{q}_r(\varepsilon)$, then it may be assumed that $\boldsymbol{q}_r(0) = \boldsymbol{e}_r$, and then for sufficiently small ε we always normalize $\boldsymbol{q}_r$ so that its rth element is 1. Define $Q = [\boldsymbol{q}_1, \boldsymbol{q}_2, \ldots, \boldsymbol{q}_n]$, and we then have

$$Q(\varepsilon) = I + \varepsilon R + \varepsilon^2 S + \cdots \tag{9.5.5}$$

where *the diagonal elements* of R, S, ... *are all zero*. The equations

$$(I\lambda_r^2 + \varepsilon\beta\lambda_r + W^2)\, \boldsymbol{q}_r = \boldsymbol{0}, \qquad r = 1, 2, \ldots, n$$

may now be written:

$$Q\Lambda^2 + \varepsilon\beta Q\Lambda - \Lambda^2 Q = 0, \tag{9.5.6}$$

and the procedure is now to substitute in this equation the series (9.5.4) and (9.5.5), to collect the terms in powers of ε, and then equate these separately to zero. These equations then provide the unknown elements of $\Lambda_1, \Lambda_2, \ldots$ and R, S, ...

The coefficient of ε gives

$$(R\Lambda_0^2 - \Lambda_0^2 R) + 2\Lambda_0\Lambda_1 + \beta\Lambda_0 = 0. \tag{9.5.7}$$

For the moment, let us confine our attention to a latent root $\lambda_u(\varepsilon)$ for which $\lambda_{0u} \neq 0$ and ω_{u0} is an unrepeated natural frequency. Consider the uth diagonal element of the eqn. (9.5.7), remembering that Λ_0, Λ_1 are diagonal and that the diagonal elements of R are zero. It is found that

$$\lambda_{1u} = -\tfrac{1}{2}\beta_{uu}. \tag{9.5.8}$$

The vth off-diagonal element of the uth column of (9.5.7) yields

$$r_{vu} = \frac{\beta_{vu}\lambda_{0u}}{\lambda_{0v}^2 - \lambda_{0u}^2}, \qquad v \neq u, \tag{9.5.9}$$

or, using (9.5.3),

$$r_{vu} = \frac{i\beta_{vu}\omega_{uo}}{\omega_{u0}^2 - \omega_{v0}^2}, \qquad v = 1, 2, \ldots, u-1, u+1, \ldots, n. \tag{9.5.10}$$

Equations (9.5.8) and (9.5.10) determine the first order perturbations contained in $\lambda_u(\varepsilon)$ and $\boldsymbol{q}_u(\varepsilon)$. The latter equation demonstrates once more that the range of validity of the results depends on the proximity of the undamped natural frequencies ω_{v0} to ω_{u0}. That is, the dimensions of the domains in which the expansions (9.5.5) and (9.5.6) are valid are expected to approach zero if $\omega_{v0} \to \omega_{u0}$ for some $v \neq u$.

Equating the coefficient of ε^2 in (9.5.6) to zero gives

$$(S\Lambda_0^2 - \Lambda_0^2 S) + (\Lambda_1^2 + 2\Lambda_0\Lambda_2) + 2R\Lambda_0\Lambda_1 + \beta(\Lambda_1 + R\Lambda_0) = 0.$$
(9.5.11)

The uth diagonal element of this equation yields

$$(\lambda_{1u}^2 + 2\lambda_{0u}\lambda_{2u}) + \beta_{uu}\lambda_{1u} + \lambda_{0u}\sum_{w=1}^{n}\beta_{uw}r_{wu} = 0.$$
(9.5.12)

If we continue on the preceding assumptions with respect to λ_{0u}, ω_{u0} we obtain

$$\lambda_{2u} = \frac{\beta_{uu}^2}{8\lambda_{ou}} - \frac{1}{2}\sum_{w=1}^{n}\beta_{uw}r_{wu}.$$

Using (9.5.9) and assuming that β is symmetric, it is found that

$$\lambda_{2u} = -\frac{i\beta_{uu}^2}{8\omega_{u0}} - \frac{1}{2}i\omega_{u0}\sum_{\substack{w=1 \\ w \neq u}}^{n}\frac{\beta_{uw}^2}{\omega_{u0}^2 - \omega_{w0}^2}.$$
(9.5.13)

We may confirm the results (9.5.8) and (9.5.13) by comparing them with those obtained by explicit computation of the derivatives, namely, (9.4.4) and (9.4.5).

The vth element of the uth column $(v \neq u)$ of (9.5.11) gives:

$$s_{vu}(\lambda_{0u}^2 - \lambda_{0v}^2) + 2r_{vu}\lambda_{0u}\lambda_{1u} + \beta_{vu}\lambda_{1u} + \lambda_{0u}\sum_{w=1}^{n}\beta_{vw}r_{wu} = 0.$$
(9.5.14)

This equation in s_{vu} leads to the result

$$s_{vu} = \frac{1}{2}\beta_{uu}\beta_{vu}\frac{\omega_{u0}^2 + \omega_{v0}^2}{(\omega_{u0}^2 - \omega_{v0}^2)^2} - \frac{\omega_{u0}^2}{\omega_{u0}^2 - \omega_{v0}^2}\sum_{\substack{w=1 \\ w \neq u}}^{n}\frac{\beta_{vw}\beta_{wu}}{\omega_{u0}^2 - \omega_{w0}^2}.$$
(9.5.15)

Equations (9.5.13) and (9.5.15) now complete the analysis for the second order perturbation terms in $\lambda_r(\varepsilon)$ and $q_r(\varepsilon)$.

Before going on to consider the general case in which some undamped natural frequencies coincide, we should dispose of the case in which $\omega_{10} = 0$, $\omega_{20} \neq 0$ in (9.5.3). Although there is a double null latent root of (9.1.6) at $\varepsilon = 0$, there is only one null entry in the leading diagonal of Λ_0 of (9.5.4). The coefficient of ε shown in (9.5.7) now tells us nothing about λ_{11}, but (9.5.9) is still valid and reduces to

$$r_{v1} = 0, \qquad v = 2, 3, \ldots, n.$$

Equation (9.5.12) now gives

$$\lambda_{11} = -\beta_{11},$$

and (9.5.14) gives

$$s_{v1} = \frac{\beta_{11}\beta_{v1}}{\omega_{r0}^2}, \qquad v = 2, 3, \dots, n.$$

In order to find λ_{21} we must consider the coefficient of ε^3 in (9.5.7). When this is done it is found that $\lambda_{2m} = 0$. As we mentioned in § 9.4, there is one null latent root of the system at $\varepsilon = 0$ which gives rise to an identically null latent root of (9.1.6). The latent vector associated with this is also independent of ε, and so there is nothing more to be said about this root. This completes the second order perturbation analysis for any $\lambda_r(\varepsilon)$ to which there corresponds a distinct entry, λ_{0r}, in Λ_0.

9.6 The Case of Coincident Undamped Natural Frequencies

It may be argued that the possibility of coincident undamped natural frequencies arising in practice is very remote. Although this may be the case for non-zero frequencies, ω_{r0}, it often happens that more than one of these numbers is zero. For example, there arise multiple null frequencies when a mechanical system is not elastically restrained in space. Our analysis will also pave the way for an attack on the more likely, and troublesome, problem in which some natural frequencies are relatively close together—in a sense to be determined.

We assume now that $\omega_{10} = \omega_{20} = \cdots = \omega_{\alpha 0} \neq 0$, $\alpha \leqq n$, and the remaining natural frequencies are non-zero. Let us examine the structure of the matrix Q employed in § 9.1 (to reduce eqns. (9.1.1) to the form (9.1.4)) in the light of Theorem 2.2, recalling that A and C are now symmetric, so that $R_\alpha = Q_\alpha$ in that theorem. We may identify Q_α with the first α columns of Q and writing $Q_{n-\alpha}$, for the matrix of the last $n - \alpha$ columns of Q, we have

$$Q = [Q_\alpha, Q_{n-\alpha}].$$

Let G be an arbitrary $\alpha \times \alpha$ non-singular matrix; then Theorem 2.2 implies that the matrix

$$Q^* = [Q_\alpha G, Q_{n-\alpha}]$$

will also give a transformation to principal coordinates provided that it is coupled with premultiplication by

$$[G^{-1}Q'_\alpha, Q'_{n-\alpha}]$$

rather than $Q^{*\prime}$. On applying this transformation to (9.1.1) we obtain in place of (9.1.4):

$$\ddot{\boldsymbol{q}} + \varepsilon\beta^*\dot{\boldsymbol{q}} + W^2\boldsymbol{q} = \boldsymbol{0}, \tag{9.6.1}$$

where

$$\beta^* = \begin{bmatrix} G^{-1}B_{11}G & G^{-1}B_{12} \\ B'_{12}G & B_{22} \end{bmatrix} \tag{9.6.2}$$

and B_{11}, B_{12}, B_{22} are the obvious partitions of $\beta = Q'BQ$ of order $\alpha \times \alpha$, $\alpha \times (n - \alpha)$, and $(n - \alpha) \times (n - \alpha)$ respectively.

Let us return now for a moment to the analysis of § 9.5, but with the hypothesis of this discussion. What goes wrong with the former analysis? The problem arises when we attempt to evaluate r_{vu}, where $v \neq u$ and $u, v \leqq \alpha$. The u, v element of (9.5.7) would now give $\beta_{vu}\lambda_{0u} = 0$. But $\lambda_{0u} \neq 0$ and, in general, $\beta_{vu} \neq 0$ and we seem to have reached an impasse. We may circumvent the impasse, however, by using β^* of (9.6.2) rather than β, provided the $\alpha \times \alpha$ matrix $B_{11} = Q'_\alpha B Q_\alpha$ is of simple structure, and *we choose G to be a matrix such that the leading partition $G^{-1}B_{11}G$ of β^* is diagonal.* We then automatically satisfy the equation $\beta^*_{vu}\lambda_{0u} = 0$, for G is chosen so that $\beta^*_{vu} = 0$ $(v \neq u)$.

We now seek the latent roots and right latent vectors satisfying

$$(I\lambda^2 + \varepsilon\beta^*\lambda + W^2)\,\boldsymbol{q} = \boldsymbol{0},$$

and, as described in § 9.5, we apply the perturbation technique to

$$Q\Lambda^2 + \varepsilon\beta^*Q\Lambda - \Lambda_0^2 Q = 0, \tag{9.6.3}$$

using the expansions (9.5.4) and (9.5.5). As in (9.5.7) we have

$$(R\Lambda_0^2 - \Lambda_0^2 R) + 2\Lambda_0\Lambda_1 + \beta^*\Lambda_0 = 0.$$

The leading $(\alpha \times \alpha)$ submatrix of $(R\Lambda_0^2 - \Lambda_0^2 R)$ is now null, so the corresponding submatrices of this equation now give

$$G^{-1}B_{11}G + 2(\Lambda_1)_\alpha = 0.$$

Now G is chosen so that $G^{-1}B_{11}G$ is diagonal, say diag $\{\beta^*_{11}, \ldots, \beta^*_{\alpha\alpha}\}$, the diagonal elements being the eigenvalues of B_{11}. Thus,

$$\lambda_{1u} = -\tfrac{1}{2}\beta^*_{uu}, \qquad u = 1, 2, \ldots, \alpha.$$

In practice, the matrix G and the corrections λ_{1u} can be constructed as an orthonormal set of right eigenvectors of $\frac{1}{2}B_{11}$ and the corresponding eigenvalues with their signs reversed. If $\alpha = 2$, we can evaluate λ_{11} and λ_{12} explicitly as

$$\lambda_{11}, \lambda_{12} = -\tfrac{1}{4}(\beta_{11} + \beta_{22}) \pm \tfrac{1}{4}[(\beta_{11} - \beta_{22})^2 + 4\beta_{12}^2]^{1/2} \tag{9.6.4}$$

We should also note the close analogy between the assumption that B_{11} is of simple structure and the hypothesis of Theorem 2.7.

From this point on we can proceed with the perturbation method as described in the preceding section. It is found that eqns. (9.5.8) and (9.5.13) are still true *if we replace β by β^* wherever it appears.* For the elements of matrices R and S we obtain (provided that $\beta^*_{11}, \ldots, \beta^*_{\alpha\alpha}$ are distinct)

$$r_{vu} = \begin{cases} \dfrac{i\beta^*_{vu}\,\omega_{u0}}{\omega^2_{u0} - \omega^2_{v0}}, & \text{for } v > \alpha \quad \textit{or} \quad u > \alpha, \\[2ex] \dfrac{i\omega_{u0}}{\beta^*_{uu} - \beta^*_{vv}} \displaystyle\sum_{w=\alpha+1} \dfrac{\beta^*_{vw}\beta^*_{wu}}{\omega^2_{u0} - \omega^2_{w0}}, & \text{for } v \leqq \alpha \textit{ and } u \leqq \alpha, \end{cases} \tag{9.6.5}$$

$$s_{vu} = \begin{cases} \dfrac{\beta^*_{vu}\beta^*_{uu}(\omega^2_{v0} + \omega^2_{u0})}{2(\omega^2_{u0} - \omega^2_{v0})^2} + \dfrac{i\omega_{u0}}{\omega^2_{u0} - \omega^2_{v0}} \displaystyle\sum_{w=1}^{n} \beta^*_{vw} r_{wu}, & \text{for } v > \alpha \textit{ or } u > \alpha, \\[2ex] \dfrac{r_{vu}(2\lambda_{01}\lambda_{2u} - \lambda^2_{1u})}{\lambda_{01}(\beta^*_{uu} - \beta^*_{vv})} + \dfrac{1}{\beta^*_{uu} - \beta^*_{vv}} \displaystyle\sum_{w=\alpha+1}^{n} \beta^*_{vw} s_{wu}, & \text{for } v \leqq \alpha \textit{ and } u \leqq \alpha. \end{cases} \tag{9.6.6}$$

The repeated natural frequency most likely to arise in practice is zero, as we have already noted, and a separate analysis is required in this case. If we suppose that $\lambda_{0u} = 0$ for $u \leqq \alpha$ and if the roots λ_{0u} are distinct and non-zero for $\alpha < u \leqq n$, it is again found necessary to reduce the matrix $G^{-1}B_{11}G$ of (9.6.2) to diagonal form. Having done this, and examined coefficients of ε, ε^2, ε^3, and ε^4 it is found that

$$\lambda_{1u} = -\beta^*_{uu}, \qquad \lambda_{2u} = 0, \qquad u \leqq \alpha,$$

$$r_{vu} = \begin{cases} 0, & \text{for } u \leqq \alpha, \\[1ex] \dfrac{i\beta^*_{vu}\omega_{u0}}{\omega^2_{u0} - \omega^2_{v0}}, & \text{for } u > \alpha, \end{cases}$$

$$s_{vu} = \begin{cases} \dfrac{\beta^*_{uu}\beta^*_{vu}}{\omega^2_{v0}}, & \text{for } v > \alpha \textit{ and } u \leqq \alpha, \\[2ex] \dfrac{-\beta^*_{uu}}{\beta^*_{uu} - \beta^*_{vv}} \displaystyle\sum_{w=\alpha+1}^{n} \dfrac{\beta^*_{vw}\beta^*_{wu}}{\omega^2_{w0}}, & \text{for } v \leqq \alpha \textit{ and } u \leqq \alpha, \end{cases} \tag{9.6.7}$$

and (9.5.15) (with β replaced by β^*) holds for $u > \alpha$. It should also be noted that, in addition to the roots for which the perturbation formulae are given, there will be α roots of the systems which are identically zero.

We observe also that the crux of the analysis is the choice of matrix G, and that the analysis breaks down completely if the leading $\alpha \times \alpha$ partition of $Q'BQ$ is not of simple structure. The physical significance of such a case is not clear. However, when the analysis does hold, then it is the distribution of damping effects in the system which determines the free modes of vibration corresponding to a multiple undamped natural frequency (re § 7.10). This suggests that an analysis of a damped system based on crude assumptions with respect to the distribution of damping effects will give unrealistic results for the modes corresponding to the roots in question, even though these roots are distinct when $\varepsilon \neq 0$.

Let us consider a simple numerical example of a problem with coincident non-zero undamped natural frequencies. We take

$$A = \frac{1}{64}\begin{bmatrix} 34 & -4 & -6 \\ -4 & 40 & -20 \\ -6 & -20 & 18 \end{bmatrix} \quad \text{and} \quad C = \frac{1}{64}\begin{bmatrix} 35 & -2 & -5 \\ -2 & 44 & -18 \\ -5 & -18 & 19 \end{bmatrix}.$$

The matrices Q and Λ_0 are then

$$Q = \begin{bmatrix} 1 & 1/\sqrt{2} & 1 \\ 0 & -1/\sqrt{2} & 2 \\ -1 & 1/\sqrt{2} & 3 \end{bmatrix}, \qquad \Lambda_0 = \begin{bmatrix} i & 0 & 0 \\ 0 & i & 0 \\ 0 & 0 & i\sqrt{2} \end{bmatrix}.$$

It is easily verified that $Q'AQ = I$, $Q'CQ = -\Lambda_0^2 = W^2$. Note also that any linear combination of the first two columns of Q is also a latent vector of $A\lambda^2 + C$ with latent root i.

The matrix B of (9.1.1) is chosen to be diagonal, the diagonal elements being the harmonic means of the corresponding elements of A and C. The symmetric matrix β of (9.1.6) is then

$$\begin{bmatrix} 0.827963 & 0.176811 & -0.327865 \\ & 0.741734 & 0.067080 \\ & & 5.761640 \end{bmatrix}.$$

The eigenvalues of the leading 2×2 partition of β are computed (or formula (9.6.4) is applied); dividing by two and reversing the sign gives

$$\lambda_{11} = -0.301428, \qquad \lambda_{12} = -0.483420.$$

The corresponding eigenvectors yield the orthogonal matrix

$$G = \begin{bmatrix} 0.617696 & 0.786417 \\ -0.786417 & 0.617696 \end{bmatrix}.$$

We now compute β^* using (9.6.2):

$$\beta^* = \begin{bmatrix} 0.602856 & 0 & -0.255274 \\ & 0.966840 & -0.216403 \\ & & 5.761640 \end{bmatrix}.$$

The preferred set of natural modes of vibration (§ 7.3) for the undamped system is given by

$$Q^* = \begin{bmatrix} 0.061616 & 1.223194 & 1 \\ 0.556081 & -0.436777 & 2 \\ -1.173777 & -0.349639 & 3 \end{bmatrix}.$$

We now apply the results obtained above. The following values for the latent roots are obtained when $\varepsilon = 0.01$. The last row of the table gives the roots calculated exactly to seven decimal places for comparison.

	$i = 1$		$i = 2$		$i = 3$	
	Real	Imaginary	Real	Imaginary	Real	Imaginary
λ_{0i}	0	1.0	0	1.0	0	1.4142 136
λ_{1i}	−0.0030 143	1.0	−0.0048 342	1.0	−0.0288 082	1.4142 136
λ_{2i}	−0.0030 143	0.9999 987	−0.0048 342	0.9999 907	−0.0288 082	1.4139 122
λ_i	−0.0030 165	0.9999 972	−0.0048 317	0.9999 922	−0.0288 084	1.4139 122

The first of the two columns of complex numbers given below is the last column of $(I + \varepsilon R + \varepsilon^2 S)$ of (9.5.5). The second column of this table gives the exact latent vector for comparison.

$(I + \epsilon R + \epsilon^2 S)$		Exact solution	
Real	Imaginary	Real	Imaginary
−0.0001 898	−0.0036 101	−0.0001 894	−0.0036 010
−0.0001 452	−0.0030 604	−0.0001 449	−0.0030 542
1.0	0	1.0	0

9.7 The Case of Neighboring Undamped Natural Frequencies

When several undamped natural frequencies are clustered together—that is, when the separation between any two of the cluster is very much less than that between one of the cluster and any one not in the cluster—the system has confusing resonance properties, as in the case of coincident natural frequencies, but does not permit us to assign the matrix G (introduced in the latter case) in such a way that the distribution of damping will isolate appropriate natural modes. However, that line of attack may be modified slightly as follows. We first consider the undamped system, for which we have the latent root problem:

$$(A\lambda^2 + C)\,\boldsymbol{p} = \boldsymbol{0}, \tag{9.7.1}$$

and construct a "neighboring" system in which all the natural frequencies not in the cluster are reproduced exactly, but in which the clustered frequencies are replaced by a multiple natural frequency. The natural modes are the same for both systems. We fortunately have theoretical solutions to this problem in § 2.9. It is shown there that the system we seek may be produced by a perturbation of either A or C. We choose to consider a perturbation of A, and the technique is now to view (9.1.6) as being simultaneously subject to two perturbations: one is of A, and the other is the damping effects, as considered previously. The unperturbed system now has coincident natural frequencies, and the matrix G is once more at our disposal. The combination of perturbations then determines G, and the classical techniques can be applied once more.

Suppose, then, that at $\varepsilon = 0$ we have a cluster of α undamped natural frequencies (not clustered about the null natural frequency), and assign these to be $\omega_{10}, \omega_{20}, \ldots, \omega_{\alpha 0}$. (Note that the ascending order of magnitude is now abandoned.) For simplicity, we also assume that $\omega_{r0} \neq 0$, $r = 1, 2, \ldots, n$. Let ω_{m0}^2 be a convenient mean value of the clustered natural frequencies. We assume that

$$|\omega_{r0}^2 - \omega_{m0}^2| \ll \omega_{m0}^2, \qquad r = 1, 2, \cdots, \alpha \tag{9.7.2}$$

and write

$$\left.\begin{aligned} W^2 &= \operatorname{diag}\{w_{10}^2, \cdots, w_{\alpha 0}^2, \omega_{\alpha+1,0}^2, \cdots, \omega_{n0}^2\}, \\ \mathrm{W}_0^2 &= \operatorname{diag}\{\omega_{m0}^2, \cdots, \omega_{m0}^2, \omega_{\alpha+1,0}^2, \cdots, \omega_{n0}^2\}. \end{aligned}\right\} \tag{9.7.3}$$

For the perturbed system whose latent roots are the diagonal elements of W_0^2 with the signs reversed we consider the problem

$$(A_1\lambda^2 + C)\,\boldsymbol{q} = \boldsymbol{0}$$

assuming that the latent vectors are those of (9.7.1) normalized in such a way that

$$Q'A_1Q = I, \qquad Q'CQ = W_0^2.$$

There now exists a diagonal matrix Δ such that

$$Q'AQ = I - \Delta \qquad \text{and} \qquad Q'CQ = (I - \Delta)\,W^2.$$

If we also write

$$W^2 = W_0^2 + W_1, \tag{9.7.4}$$

then since

$$(I - \Delta)\,W^2 = W_0^2,$$

we have

$$\Delta = W_1(W^2)^{-1}. \tag{9.7.5}$$

Making the transformation of (9.1.1) to principal coordinates defined by the above matrix Q, we arrive at the problem:

$$\{(I - \Delta)\,\lambda^2 + \varepsilon\beta\lambda + \mathrm{W}_0^2\}\,\boldsymbol{q} = \boldsymbol{0}. \tag{9.7.6}$$

Note that we do not have to calculate A_1 explicitly. We simply calculate Δ from (9.7.5), and Δ will be small depending on the degree to which (9.7.2) is satisfied. In order to reduce the problem to the type considered in § 9.6, we now suppose that the magnitudes of the two perturbations involved are of the same order, and we replace Δ in (9.7.6) by εD, it being assumed that the first α diagonal elements of D are thereby $0(1)$ as $\varepsilon \to 0$. The system obtained when $\varepsilon = 0$ then has coincident natural frequencies, and we may introduce the matrix G into Q as described in § 9.6. The resulting equation is then

$$\{(I - \varepsilon D^*)\,\lambda^2 + \varepsilon\beta^*\lambda + W_0^2\}\,\boldsymbol{q} = \boldsymbol{0},$$

where β^* is given by (9.6.2) and

$$D^* = \begin{bmatrix} G^{-1}D_{11}G & 0 \\ 0 & 0 \end{bmatrix}, \tag{9.7.7}$$

D_{11} being the leading $\alpha \times \alpha$ partition of D. Notice that, even though D_{11} is diagonal, $G^{-1}D_{11}G$ is not necessarily so.

We now write, as in equation (9.6.3),

$$(I - \varepsilon D^*)\,Q\Lambda^2 + \varepsilon\beta^*Q\Lambda - \Lambda_0^2Q = 0,$$

and substitute the expansions (9.5.4) and (9.5.5). The first $\alpha \times \alpha$ partition of the coefficient of ε gives

$$G^{-1}(B_{11} - iD_{11}\omega_{m0})\,G + 2(\Lambda_1)_\alpha = 0.$$

On the assumption that $(B_{11} - iD_{11}\omega_{m0})$ is of simple structure we can now determine G so that $G^{-1}(B_{11} - iD_{11}\omega_{m0})\, G$ is diagonal. The diagonal elements of $(\Lambda_1)_\alpha$ may then be determined. In practice, we compute these numbers λ_{1u}, $u = 1, 2, \ldots, \alpha$ as the eigenvalues of $(B_{11} - iD_{11}\omega_{m0})$ (which must later be assumed to be distinct), and G is a matrix of independent eigenvectors, one for each eigenvalue. In principle this is no more difficult than the computation met with in § 9.6, but the computation is complicated by the fact that the matrix we investigate is now complex, where formerly it was real.

It is left to the reader either to obtain the rather complicated formulae for Λ_2, R, and S by the now familiar method, or to refer to the original paper (Lancaster [1]) for further details. We illustrate the technique with a numerical example.

Let

$$A = \frac{1}{63.9444}\begin{bmatrix} 33.6596 & -4.5208 & 5.5404 \\ & 40.5584 & -20.2008 \\ & & 17.9796 \end{bmatrix}$$

and let C be the matrix C of the example in § 9.6. If is found that

$$\omega_{10}^2 = 1.02, \qquad \omega_{20}^2 = 0.98, \qquad \omega_{30}^2 = 2.00,$$

exactly, and the latent vectors are those of the problem in § 9.6. We choose $\omega_{m0}^2 = 1.00$, so that in (9.7.3) and (9.7.5) we have

$$W_0^2 = \begin{bmatrix} 1 & 0 & 0 \\ 0 & 1 & 0 \\ 0 & 0 & 2 \end{bmatrix}, \qquad \Delta = \begin{bmatrix} \dfrac{0.02}{1.02} & 0 & 0 \\ 0 & \dfrac{-0.02}{0.98} & 0 \\ 0 & 0 & 0 \end{bmatrix}$$

We use the same damping matrix, B, employed in § 9.6 and we again take $\varepsilon = 0.01$. Comparing $\varepsilon\beta\omega$ with $\Delta\omega^2$, we observe that the perturbations are of the same order of magnitude and write

$$\Delta = \varepsilon D = (0.01)\begin{bmatrix} \dfrac{2}{1.02} & 0 & 0 \\ 0 & \dfrac{-2}{0.98} & 0 \\ 0 & 0 & 0 \end{bmatrix}.$$

The matrix $B_{11} - iD_{11}\omega_{m0}$ is now (to six decimal places):

$$\begin{bmatrix} 0.827963 - i(1.960784) & 0.176811 \\ 0.176811 & 0.741734 - i(2.040816) \end{bmatrix}.$$

We compute the eigenvalues, divide by two, and reverse the sign to obtain

$$\lambda_{11} = -0.44066 + i(0.976480),$$
$$\lambda_{12} = -0.370782 - i(1.016496),$$
$$G = \begin{bmatrix} -1.000980 + i(0.000042) & 0.000960 + i(0.044295) \\ -0.000960 - i(0.044295) & -1.000980 + i(0.000042) \end{bmatrix}.$$

We may now compute β^* and D^*, and the rest is purely manipulative.

The tables given below are comparable to those of § 9.6. The first table compares the successive approximations to the latent roots with their exact values, and the second compares the last column of $(I + \varepsilon R + \varepsilon^2 S)$ with the exact latent vector.

	$i = 1$		$i = 2$		$i = 3$	
	Real	Imaginary	Real	Imaginary	Real	Imaginary
λ_{0i}	0	1.0	0	1.0	0	1.4142 136
λ_{1i}	−0.0041 407	1.0097 648	−0.0037 078	0.9898 350	−0.0288 082	1.4142 136
λ_{2i}	−0.0042 217	1.0099 054	−0.0036 332	0.9899 841	−0.0288 082	1.4139 122
λ_i	−0.0042 230	1.0099 079	−0.0036 339	0,9899 817	−0,0288 085	1.4139 119

$(I + \varepsilon R + \varepsilon^2 S)_3$		Exact vector	
Real	Imaginary	Real	Imaginary
0.0002 721	0.0048 249	0.0002 875	0.0048 152
0.0001 534	−0.0009 257	0.0001 517	−0.0009 245
1.0	0	1.0	0

The leading entries in the first of these tables may be a little misleading. The values of 1.0 for $\mathscr{I}(\lambda_{01})$ and $\mathscr{I}(\lambda_{02})$ are, of course, purposely removed from the undamped natural frequencies, ω_{10} and ω_{20}; i.e., 1.0099505 and 0.9899495 respectively.

Finally, let us compare the results for λ_1 obtained by the present method with those obtained from eqns. (9.5.8) and (9.5.13), which were derived on the basis of distinct natural frequencies. These two formulae give rise to the following table. Thus, the analysis of this

	$i = 1$	
	Real	Imaginary
λ_{0i}	0	1.0099505
λ_{1i}	−0.0041398	1.0099505
λ_{2i}	−0.0041398	1.0098915

section gives improved accuracy in the second order analysis, and this improvement could be expected to be more marked if the natural frequencies were closer than in the example we have used. It should also be noted that, if we only require a first order correction to the latent roots, then the analysis of § 9.5 appears to be quite as good as this more complicated one. This is to be expected, as (9.5.8) is not affected by clustered natural frequencies.

BIBLIOGRAPHICAL NOTES

These notes are intended to supplement those made in the body of the monograph and are collected here in order to reduce the number of discontinuities in the next.

The book with objectives most closely related to our own is probably that of Frazer *et al.* [1], which contains may results closely related to or overlapping with, those of Chapters 1–4, 6, and 7. The voluminous *Mechanics of Vibration* by Bishop and Johnson [1] also covers much of the ground covered in our later chapters, but in a much more leisurely fashion, with more attention to physical detail, and without the invaluable aid of matrices. For an elegant mathematical treatment of vibration problems, the reader should consult Gantmacher and Krein [1].

Our first chapter contains no results which will be unfamiliar to those conversant with the theory of matrices. The only unfamilar feature is likely to be the usage of the terms "eigenvalue" and "latent root." The distinction is most important for later developments. It was first adopted by Lancaster [3] and has since proved very useful. It would be a pity if the more expressive term "latent root" were to go out of use altogether. Theorems 2.5, 2.6 and 2.7 have not previously appeared in book form (Lancaster [5]), and provide useful extensions for the classical results of Rellich [1].

The proof of Theorem 3.5 on the decomposition of regular λ-matrices of degree two is new. Different proofs have been given by Lancaster [2], when the theorem was first published, and by Ostrowski [4]. Closely related results in a more general context have been published by Krein and Langer [1]. For an account of results related to Theorem 3.7 see Chapter 8 of MacDuffee [1].

The key to the proof of the vital Theorem 4.3, concerning the inverse of a simple λ-matrix, is the discovery of the related matrix pencil of (4.2.8). This appears to have originated with Foss [1]. Lancaster first proved the theorem in its present form in his reference [2], though it was published earlier and independently by Guderley [1]. Theorem 4.8 facilitates comparison of our results with those of Frazer *et al.*, who develop these results in terms of the adjoint matrix and its derivatives. Theorem 4.10 is due to Ostrowski [4]. The generalized Rayleigh quotient of Theorem 4.11 was obtained by Lancaster [3] after being stimulated by Ostrowski's papers [2].

The most useful algorithms of Chapter 5 depend on the trace theorem, Theorem 5.1. They are of very recent origin (Lancaster [6]). Before undertaking large computations with the algorithms developed here, the inexperienced computer is advised to absorb some of the technical know-how of Wilkinson [1], [2], or Durand [1], for example. Contributions to the theory of computation with equivalent linear problems have been made by Bauer [1] and Dimsdale [1]. The "fixed point" theorem of § 5.6 is easily proved in a much wider context. See, for example, works by Collatz [1], [2], or Chapter 14 of Todd [1], for proofs and a variety of applications. Urabe [1], [2], has investigated the sources and magnitudes of the oscillatory errors referred to in § 5.8. That is, the errors whose magnitudes determine a lower bound for the cut-off parameter, ε. Stability problems of the kind discussed in § 5.9 have arisen in aircraft flutter problems (Frazer *et al.* [1]), control problems (Tarnove [1]), and hydrodynamics (Dolph and Lewis [1]).

A careful examination of basic concepts in linear vibration problems and a classification of stability problems can be found in the paper by Ziegler [1]. More accurate mathematical descriptions of the mechanism of internal damping than those of § 7.2 have been attempted by Milne [2] and Caughey [1]. Milne [1] has also discussed the effects of approximating a given system by one with fewer degrees of freedom. Most of the results of § 7.6, for overdamped systems originate with Duffin [1]. However, his analysis goes further than ours, and he develops an elegant extension of the variational method which demonstrates the effect of imposing constraints on the system. The variational method does not appear to have any application to the general damped systems, and so we have not followed up these ideas. Duffin [2] has also discussed the gyroscopic systems of § 7.7. In the case of a conservative holonomic system the Hamiltonian technique can be used to obtain a system of first order differential equations. Erdös [1] has followed this line of attack and gives an explicit solution of the resulting equations in the case of a simple matrix of coefficients. A good summary of vibration theory for systems with several degrees of freedom is given by Crandall and McCalley [1], whose treatment follows lines very like our own, and includes an account of some more familiar numerical methods.

Following the original presentation by Fraeijs de Veubeke [1] of the method of stationary phase, his technique (among others) has been investigated at some length by Bishop and Gladwell [1]. However, though they do not say so, their analysis is only valid when applied to positive definite stiffness and damping matrices. The course of Traill-Nash's practical applications of the technique, [1], [2], is interesting. For further discussion of pencils $A\lambda + C$ where A and C are both non-negative definite, see the note by Newcomb [1] and Chapter 12 of Gantmacher [2].

The results of § 9.2 were first obtained by Falk [1], who describes his results in an elegant graphical manner. The methods of § 9.3 are due to Lancaster [4], and largely owe their existence to the stimulus received by this author on noting Müller's transformation [1]. The perturbation analyses of § 9.4–§ 9.7 originate with Lord Rayleigh [1], but were extended by Lancaster [1] to the problems of coincident and clustered natural frequencies. The treatment of problems with coincident natural frequencies followed naturally from the treatment of Courant and Hilbert [1], p. 346.

REFERENCES

AITKEN, A. C. [1] *Determinants and Matrices*, Oliver and Boyd, 1944.

ASHER, G. W. [1] A method of normal mode excitation utilizing admittance measurements *Proc. Inst. Aero. Sci.*, "Dynamics and Aero-elasticity," 1958, 69–76.

BAUER, F. L. [1] Zur numerischen Behandlung von algebraischen Eigenwertproblemen höherer Ordnung, *Zeit. angew. Math. u. Mech.* **36**, 245–246 (1956).

BAUER, F. L., and HOUSEHOLDER, A. S. [1] Absolute norms and characteristic roots, *Num. Math.* **3**, 241–246 (1961).

BELLMAN, R., [1] *Introduction to Matrix Analysis*, McGraw Hill, 1960.

BISHOP, R. E. D., and GLADWELL, G. M. L. [1] An investigation into the theory of resonance testing, *Phil. Trans. Roy. Soc.* **225**, 241–280 (1963).

BISHOP, R. E. D., and JOHNSON, D. C. [1] *The Mechanics of Vibration*, Cambridge University Press, 1960.

BRAUER, A. [1] Limits for the characteristic roots of a matrix, II, *Duke Math. J.* **14**, 21–26 (1947).

CARSLAW, H. S., and JAEGAR, J. C. [1] *Operational Methods in Applied Mathematics*, Oxford University Press, 1953.

CAUGHEY, T. K. [1] Vibration of dynamic systems with linear hysteretic damping, *Proc. 4th U.S. Nat. Congr. Appl. Mechanics*, 1962, 87–97.

CODDINGTON, E. A., and LEVINSON, N. [1] *Theory of Ordinary Differential Equations*, McGraw-Hill, 1955,

COLLATZ, L. [1] Einige Anwendungen Funktionalanalytischer Methoden in der praktischen Analysis, *Zeit. angew. Math. u. Phys.* **4**, 327–357 (1953); [2] *Funktionalanalysis und numerische Mathematik*, Springer, 1964.

COURANT, R., and HILBERT D. [1] *Methods of Mathematical Physics, I*, Interscience, 1953.

CRANDALL, S. H. [1] Iterative procedures related to the relaxation methods for eigenvalue problems, *Proc. Roy. Soc. London* A, **207**, 416–423 (1951); [2] (editor) *Random Vibration*, Wiley, 1958.

CRANDALL, S. H., and MCCALLEY, R. B., JR. [1] Numerical methods of analysis, ch. 28, *Shock and Vibration Handbook*, vol. 2, McGraw-Hill, 1961.

DIEDERICH, F. W. [1] The dynamic response of a large airplane to continuous random atmospheric turbulence, *J. Aero. Sci.* **23**, 917–930 (1956).

DIMSDALE, B. [1] Characteristic roots of a matric polynomial, *J. Soc. Ind. Appl. Math.* **8**, 218–223 (1960).

DOLPH, C. L., and LEWIS, D. C. [1] On the application of infinite systems of ordinary differential equations to perturbations of plane Poiseuille flow, *Quart. Appl. Math.* **16**, 97–110 (1958).

DUFFIN, R. J. [1] A minimax theory for overdamped networks, *J. Rat. Mech.*

Anal. **4**, 221–233 (1955); [2] The Rayleigh-Ritz method for dissipative or gyroscopic systems, *Quart. Appl. Math.* **18**, 215–221 (1960).

DURAND, E. [1] *Solutions numériques des equations algébriques*, tome 1, Masson & Cie. (Paris), 1960.

ERDÖS, P. [1] Kleine Schwingungen dynamischer Systeme, *Zeit. angew. Math. u. Phys.* **4**, 215–219 (1953).

FALK, S. [1] Klassifikation gedämpfter Schwingungssysteme und Eingrenzung ihrer Eigenwerte, *Ing.-Arch.* **29**, 436–444 (1960).

FATOU, P. [1] Sur l'itération des fonctions transcendantes entières, *Acta Math.* **47**, 337–370 (1926).

FERRAR, W. L. [1] *Finite Matrices*, Oxford University Press, 1951.

FLOOD, M. M. [1] Division by non-singular matric polynomials, *Ann. Math.* **36**, 859–869 (1935).

FOSS, K. A. [1] Coordinates which uncouple the equations of motion of damped linear dynamic systems, *J. Appl. Mech.* **25**E, 361–364 (1958).

FRAEIJS DE VEUBEKE, B. M. [1] A variational approach to pure mode excitation based on characteristic phase lag theory, *AGARD Report* 39, 1956.

FRAZER, R. A., DUNCAN, W. J., and COLLAR, A. R. [1] *Elementary Matrices*, Cambridge University Press, 1955.

GANTMACHER, F. R. [1] *The Theory of Matrices*, vol. 1, Chelsea, 1959; [2] *The Theory of Matrices*, vol. 2, Chelsea, 1959.

GANTMACHER, F. R., and KREIN, M. G. [1] *Oszillationsmatrizen, Oszillationskerne und kleine Schwingungen mechanischer Systeme*, Akademie-Verlag, Berlin, 1960.

GOULD, S. H. [1] *Variational Methods for Eigenvalue Problems*, University of Toronto, 1957.

GOURSAT, E. [1] *Functions of a Complex Variable*, **2**, pt. 1, Ginn, 1916.

GUDERLEY, K. G. [1] On nonlinear eigenvalue problems for matrices. *J. Soc. Ind. Appl. Math.* **6**, 335–353 (1958).

HOUSEHOLDER, A. S. [1] *The Theory of Matrices in Numerical Analysis*, Blaisdell, 1964.

HOWLAND, J. L. [1] A method for computing the real roots of determinantal equations, *Am. Math. Monthly* **68**, 235–239 (1961).

JULIA, G. [1] Mémoire sur l'iteration des fonctions rationnelles, *J. math. pures et Appl.* (ser. 8) **1**, 47–245 (1918).

KREIN, M. G., and LANGER, G. K. [1] A contribution to the theory of quadratic pencils of self-adjoint operators, Soviet Math. **5**, 266–269 (1964).

LANCASTER, P. [1] Free vibrations of lightly damped systems by perturbation methods, *Quart. J. Mech. Appl. Math.* **13**, 138–155 (1960); [2] Inversion of lambda matrices and application to the theory of linear vibrations, *Arch. Rat. Mech. Anal.* **6**, 105–114 (1960); [3] A generalised Rayleigh-quotient iteration for lambda-matrices, *Arch. Rat. Mech. Anal.* **8**, 309–322 (1961); [4] Bounds for latent roots in damped vibration problems, *SIAM Rev.* **6**, 121–125 (1964); [5] On eigenvalues of matrices dependent on a parameter, *Num. Math.* **6**, 377–387 (1964); [6] Algorithms for lambda-matrices, *Num. Math.* **6**, 388–394 (1964).

LAZAN, B. J. [1] A study with new equipment of the effects of fatigue stress on damping capacity and elasticity of mild steel, *Trans. Am. Soc. Metals* **4**, 499–558 (1950).

MacDuffee, C. C. [1] *The Theory of Matrices*, Chelsea, 1946.

Milne, R. D. [1] On the estimation of structural damping from resonance tests, *J. Aero. Sp. Sci.* **27**, 339–343 (1960); [2] The steady-state response of continuous dissipative systems with small nonlinearity, I and II, *Quart. J. Mech. Appl. Math.* **14**, 229–256 (1961).

Mirsky, L. [1] *An Introduction to Linear Algebra*, Oxford University Press, 1955.

Muller, D. E. [1] A method for solving algebraic equations using an automatic computer, *Math. Tab.* **10**, 208–215 (1956).

Müller, P. H. [1] Eigenwertabschätzungen für Gleichungen vom Typ $(\lambda^2 I - \lambda A - B)\,x = 0$, *Arch. Math.* **12**, 307–310 (1961).

Newcomb, R. W. [1] On the simultaneous diagonalization of two semi-definite matrices, *Quart. Appl. Math.* **19**, 144–146 (1961).

Ostrowski, A. M. [1] Über Normen von Matrizen, *Math. Z.* **63**, 2–18 (1955); [2] On the convergence of the Rayleigh quotient interation for the computation of characteristic roots and vectors, *Arch. Rat. Mech. Anal.* pt. I, **1**, 233–241 (1957); pt. II, **2**, 423–428 (1958); pt. III, **3**, 325–340 (1959); pt. IV, **3**, 341–347 (1959); pt. V, **3**, 472–481 (1959); pt. VI, **4**, 153–165 (1960); [3] *Solution of Equations and Systems of Equations*, Academic Press, 1960; [4] On Lancaster's decomposition of a differential matricial operator, *Arch. Rat. Mech. Anal.* **8**, 238–241 (1961).

Parlett, B. [1] Laguerre's method applied to the matrix eigenvalue problem, *Math. Comp.* **18**, 464–485 (1964).

Perlis, S. [1] *Theory of Matrices*, Addison-Wesley, 1952.

Pipes, L. A. [1] *Matrix Methods for Engineering*, Prentice-Hall, 1963.

Rayleigh, Lord [1] *Theory of Sound*, vol. 1, Macmillan, 1929 ;Dover,1945.

Rellich, F. [1] Störungstheorie der Spektralzerlegung, IV, *Math. Ann.* **117**, 356–382 (1940).

Robertson, J. M., and Yorgiadis, A. J. [1] Internal friction in engineering materials, *J. Appl. Mech.* **13**, 173 (1946).

Rutishauser, H. [1] Zur Matrizeninversion nach Gauss–Jordan, *Z. Angew. Math. u. Phys.* **10**, 281–291 (1959).

Stone, B. J. [1] Best possible ratios of certain matrix norms, *Num. Math.* **4**, 114–116 (1962).

Tarnove, I. [1] Determination of eigenvalues of matrices having polynomial elements, *J. Soc. Ind. Appl. Math.* **6**, 163–171 (1958).

Taussky, O., and Zassenhaus, H. [1] On the similarity transformation between a matrix and its transpose, *Pac. J. Math.* **9**, 893–896 (1959).

Temple, G. [1] The accuracy of Rayleigh's method for calculating the natural frequencies of vibrating systems, *Proc. Roy. Soc. London* A **211**, 204–224 (1952).

Todd, J. [1] *Survey of Numerical Analysis*, McGraw-Hill, 1962.

Traill-Nash, R. W. [1] On the excitation of pure natural modes in aircraft resonance testing, *J. Aero. Sp. Sci.* **25**, 775–778 (1958); [2] Some theoretical aspects of resonance testing and proposals for a technique combining experiment and computation, *Structures and Materials Report* 280, Aeronautical Research Laboratories, Melbourne, 1961.

Turnbull, H. W., and Aitken, A. C. [1] *Canonical Matrices*, Blackie, 1952; Dover, 1961.

URABE, M. [1] Convergence of numerical iteration in solution of equations, *J. Sci. Hiroshima Univ.* **19**, 479–489 (1956); [2] Error estimation in numerical solution of equations by iteration process. *J. Sci. Hiroshima Univ.* **26**, 77–91 (1962).

WHITTAKER, E. T. [1] *Analytical Dynamics*, Cambridge University Press, 1952.

WILKINSON, J. H. [1] The evaluation of the zeros of ill-conditioned polynomials, pts. I and II, *Num. Math.* **1**, 150–180 (1959); [2] *Rounding Errors in Algebraic Processes*, Notes on Applied Science no. 32, H.M.S.O., London, 1963; [3] *The Algebraic Eigenvalue Problem*, Oxford, 1965.

ZIEGLER, H. [1] Linear elastic stability, *Zeit. angew. Math. u. Phys.* **4**, 89–121 (1953).

INDEX

A CATALOG OF SELECTED

DOVER BOOKS

IN SCIENCE AND MATHEMATICS

A CATALOG OF SELECTED

DOVER BOOKS

IN SCIENCE AND MATHEMATICS

Astronomy

BURNHAM'S CELESTIAL HANDBOOK, Robert Burnham, Jr. Thorough guide to the stars beyond our solar system. Exhaustive treatment. Alphabetical by constellation: Andromeda to Cetus in Vol. 1; Chamaeleon to Orion in Vol. 2; and Pavo to Vulpecula in Vol. 3. Hundreds of illustrations. Index in Vol. 3. 2,000pp. 6⅛ x 9¼.
23567-X, 23568-8, 23673-0 Three-vol. set

THE EXTRATERRESTRIAL LIFE DEBATE, 1750–1900, Michael J. Crowe. First detailed, scholarly study in English of the many ideas that developed from 1750 to 1900 regarding the existence of intelligent extraterrestrial life. Examines ideas of Kant, Herschel, Voltaire, Percival Lowell, many other scientists and thinkers. 16 illustrations. 704pp. 5⅜ x 8½. 40675-X

A HISTORY OF ASTRONOMY, A. Pannekoek. Well-balanced, carefully reasoned study covers such topics as Ptolemaic theory, work of Copernicus, Kepler, Newton, Eddington's work on stars, much more. Illustrated. References. 521pp. 5⅜ x 8½.
65994-1

AMATEUR ASTRONOMER'S HANDBOOK, J. B. Sidgwick. Timeless, comprehensive coverage of telescopes, mirrors, lenses, mountings, telescope drives, micrometers, spectroscopes, more. 189 illustrations. 576pp. 5⅝ x 8¼. (Available in U.S. only.)
24034-7

STARS AND RELATIVITY, Ya. B. Zel'dovich and I. D. Novikov. Vol. 1 of *Relativistic Astrophysics* by famed Russian scientists. General relativity, properties of matter under astrophysical conditions, stars, and stellar systems. Deep physical insights, clear presentation. 1971 edition. References. 544pp. 5⅝ x 8¼. 69424-0

Chemistry

CHEMICAL MAGIC, Leonard A. Ford. Second Edition, Revised by E. Winston Grundmeier. Over 100 unusual stunts demonstrating cold fire, dust explosions, much more. Text explains scientific principles and stresses safety precautions. 128pp. 5⅜ x 8½. 67628-5

THE DEVELOPMENT OF MODERN CHEMISTRY, Aaron J. Ihde. Authoritative history of chemistry from ancient Greek theory to 20th-century innovation. Covers major chemists and their discoveries. 209 illustrations. 14 tables. Bibliographies. Indices. Appendices. 851pp. 5⅜ x 8½. 64235-6

CATALYSIS IN CHEMISTRY AND ENZYMOLOGY, William P. Jencks. Exceptionally clear coverage of mechanisms for catalysis, forces in aqueous solution, carbonyl- and acyl-group reactions, practical kinetics, more. 864pp. 5⅜ x 8½.
65460-5

THE HISTORICAL BACKGROUND OF CHEMISTRY, Henry M. Leicester. Evolution of ideas, not individual biography. Concentrates on formulation of a coherent set of chemical laws. 260pp. 5⅜ x 8½. 61053-5

A SHORT HISTORY OF CHEMISTRY, J. R. Partington. Classic exposition explores origins of chemistry, alchemy, early medical chemistry, nature of atmosphere, theory of valency, laws and structure of atomic theory, much more. 428pp. 5⅜ x 8½. (Available in U.S. only.) 65977-1

GENERAL CHEMISTRY, Linus Pauling. Revised 3rd edition of classic first-year text by Nobel laureate. Atomic and molecular structure, quantum mechanics, statistical mechanics, thermodynamics correlated with descriptive chemistry. Problems. 992pp. 5⅜ x 8½. 65622-5

Engineering

DE RE METALLICA, Georgius Agricola. The famous Hoover translation of greatest treatise on technological chemistry, engineering, geology, mining of early modern times (1556). All 289 original woodcuts. 638pp. 6¾ x 11. 60006-8

FUNDAMENTALS OF ASTRODYNAMICS, Roger Bate et al. Modern approach developed by U.S. Air Force Academy. Designed as a first course. Problems, exercises. Numerous illustrations. 455pp. 5⅜ x 8½. 60061-0

DYNAMICS OF FLUIDS IN POROUS MEDIA, Jacob Bear. For advanced students of ground water hydrology, soil mechanics and physics, drainage and irrigation engineering and more. 335 illustrations. Exercises, with answers. 784pp. 6⅛ x 9¼. 65675-6

ANALYTICAL MECHANICS OF GEARS, Earle Buckingham. Indispensable reference for modern gear manufacture covers conjugate gear-tooth action, gear-tooth profiles of various gears, many other topics. 263 figures. 102 tables. 546pp. 5⅜ x 8½. 65712-4

MECHANICS, J. P. Den Hartog. A classic introductory text or refresher. Hundreds of applications and design problems illuminate fundamentals of trusses, loaded beams and cables, etc. 334 answered problems. 462pp. 5⅜ x 8½. 60754-2

MECHANICAL VIBRATIONS, J. P. Den Hartog. Classic textbook offers lucid explanations and illustrative models, applying theories of vibrations to a variety of practical industrial engineering problems. Numerous figures. 233 problems, solutions. Appendix. Index. Preface. 436pp. 5⅜ x 8½. 64785-4

STRENGTH OF MATERIALS, J. P. Den Hartog. Full, clear treatment of basic material (tension, torsion, bending, etc.) plus advanced material on engineering methods, applications. 350 answered problems. 323pp. 5⅜ x 8½. 60755-0

A HISTORY OF MECHANICS, René Dugas. Monumental study of mechanical principles from antiquity to quantum mechanics. Contributions of ancient Greeks, Galileo, Leonardo, Kepler, Lagrange, many others. 671pp. 5⅜ x 8½. 65632-2

Physics

OPTICAL RESONANCE AND TWO-LEVEL ATOMS, L. Allen and J. H. Eberly. Clear, comprehensive introduction to basic principles behind all quantum optical resonance phenomena. 53 illustrations. Preface. Index. 256pp. 5⅜ x 8½. 65533-4

ULTRASONIC ABSORPTION: An Introduction to the Theory of Sound Absorption and Dispersion in Gases, Liquids and Solids, A. B. Bhatia. Standard reference in the field provides a clear, systematically organized introductory review of fundamental concepts for advanced graduate students, research workers. Numerous diagrams. Bibliography. 440pp. 5⅜ x 8½. 64917-2

QUANTUM THEORY, David Bohm. This advanced undergraduate-level text presents the quantum theory in terms of qualitative and imaginative concepts, followed by specific applications worked out in mathematical detail. Preface. Index. 655pp. 5⅜ x 8½. 65969-0

ATOMIC PHYSICS (8th edition), Max Born. Nobel laureate's lucid treatment of kinetic theory of gases, elementary particles, nuclear atom, wave-corpuscles, atomic structure and spectral lines, much more. Over 40 appendices, bibliography. 495pp. 5⅜ x 8½. 65984-4

AN INTRODUCTION TO HAMILTONIAN OPTICS, H. A. Buchdahl. Detailed account of the Hamiltonian treatment of aberration theory in geometrical optics. Many classes of optical systems defined in terms of the symmetries they possess. Problems with detailed solutions. 1970 edition. xv + 360pp. 5⅜ x 8½. 67597-1

THIRTY YEARS THAT SHOOK PHYSICS: The Story of Quantum Theory, George Gamow. Lucid, accessible introduction to influential theory of energy and matter. Careful explanations of Dirac's anti-particles, Bohr's model of the atom, much more. 12 plates. Numerous drawings. 240pp. 5⅜ x 8½. 24895-X

ELECTRONIC STRUCTURE AND THE PROPERTIES OF SOLIDS: The Physics of the Chemical Bond, Walter A. Harrison. Innovative text offers basic understanding of the electronic structure of covalent and ionic solids, simple metals, transition metals and their compounds. Problems. 1980 edition. 582pp. 6⅛ x 9¼. 66021-4

HYDRODYNAMIC AND HYDROMAGNETIC STABILITY, S. Chandrasekhar. Lucid examination of the Rayleigh-Benard problem; clear coverage of the theory of instabilities causing convection. 704pp. 5⅜ x 8¼. 64071-X

INVESTIGATIONS ON THE THEORY OF THE BROWNIAN MOVEMENT, Albert Einstein. Five papers (1905–8) investigating dynamics of Brownian motion and evolving elementary theory. Notes by R. Fürth. 122pp. 5⅜ x 8½. 60304-0

THE PHYSICS OF WAVES, William C. Elmore and Mark A. Heald. Unique overview of classical wave theory. Acoustics, optics, electromagnetic radiation, more. Ideal as classroom text or for self-study. Problems. 477pp. 5⅜ x 8½. 64926-1

PHYSICAL PRINCIPLES OF THE QUANTUM THEORY, Werner Heisenberg. Nobel Laureate discusses quantum theory, uncertainty, wave mechanics, work of Dirac, Schroedinger, Compton, Wilson, Einstein, etc. 184pp. 5⅜ x 8½. 60113-7

ATOMIC SPECTRA AND ATOMIC STRUCTURE, Gerhard Herzberg. One of best introductions; especially for specialist in other fields. Treatment is physical rather than mathematical. 80 illustrations. 257pp. 5⅜ x 8½. 60115-3

AN INTRODUCTION TO STATISTICAL THERMODYNAMICS, Terrell L. Hill. Excellent basic text offers wide-ranging coverage of quantum statistical mechanics, systems of interacting molecules, quantum statistics, more. 523pp. 5⅜ x 8½. 65242-4

THEORETICAL PHYSICS, Georg Joos, with Ira M. Freeman. Classic overview covers essential math, mechanics, electromagnetic theory, thermodynamics, quantum mechanics, nuclear physics, other topics. First paperback edition. xxiii + 885pp. 5⅜ x 8½. 65227-0

PROBLEMS AND SOLUTIONS IN QUANTUM CHEMISTRY AND PHYSICS, Charles S. Johnson, Jr. and Lee G. Pedersen. Unusually varied problems, detailed solutions in coverage of quantum mechanics, wave mechanics, angular momentum, molecular spectroscopy, more. 280 problems plus 139 supplementary exercises. 430pp. 6½ x 9¼. 65236-X

THEORETICAL SOLID STATE PHYSICS, Vol. 1: Perfect Lattices in Equilibrium; Vol. II: Non-Equilibrium and Disorder, William Jones and Norman H. March. Monumental reference work covers fundamental theory of equilibrium properties of perfect crystalline solids, non-equilibrium properties, defects and disordered systems. Appendices. Problems. Preface. Diagrams. Index. Bibliography. Total of 1,301pp. 5⅜ x 8½. Two volumes. Vol. I: 65015-4 Vol. II: 65016-2

A TREATISE ON ELECTRICITY AND MAGNETISM, James Clerk Maxwell. Important foundation work of modern physics. Brings to final form Maxwell's theory of electromagnetism and rigorously derives his general equations of field theory. 1,084pp. 5⅜ x 8½. Two-vol. set. Vol. I: 60636-8 Vol. II: 60637-6

OPTICKS, Sir Isaac Newton. Newton's own experiments with spectroscopy, colors, lenses, reflection, refraction, etc., in language the layman can follow. Foreword by Albert Einstein. 532pp. 5⅜ x 8½. 60205-2

THEORY OF ELECTROMAGNETIC WAVE PROPAGATION, Charles Herach Papas. Graduate-level study discusses the Maxwell field equations, radiation from wire antennas, the Doppler effect and more. xiii + 244pp. 5⅜ x 8½. 65678-5

INTRODUCTION TO QUANTUM MECHANICS With Applications to Chemistry, Linus Pauling & E. Bright Wilson, Jr. Classic undergraduate text by Nobel Prize winner applies quantum mechanics to chemical and physical problems. Numerous tables and figures enhance the text. Chapter bibliographies. Appendices. Index. 468pp. 5⅜ x 8½. 64871-0